Great Feuds in Mathematics

Ten of the Liveliest Disputes Ever

HAL HELLMAN

WILEY

John Wiley & Sons, Inc.

Published by John Wiley & Sons, Inc., Hoboken, New Jersey
Published simultaneously in Canada

Design and composition by Navta Associates, Inc.

For general information about our other products and services, please contact our Customer Care Department within the United States at (800) 762-2974, outside the United States at (317) 572-3993 or fax (317) 572-4002.

Wiley also publishes its books in a variety of electronic formats. Some content that appears in print may not be available in electronic books. For more information about Wiley products, visit our web site at www.wiley.com.

Library of Congress Cataloging-in-Publication Data:

Hellman, Hal, date.
  Great feuds in mathematics : ten of the liveliest disputes ever / Hal Hellman.
        p. cm.
  Includes bibliographical references and index.
  ISBN-13 978-0-471-64877-2 (cloth)
  ISBN-10 0-471-64877-9 (cloth)
  1. Mathematics—History. 2. Mathematicians—History. I. Title.
  QA21.H45 2006
  510—dc22
                                                                2005031916

Printed in the United States of America

10 9 8 7 6 5 4 3 2 1

Contents

Acknowledgments

A book like this required that I sort through an enormous amount of material. The Internet was convenient, quick, slow, amazing, and frustrating, not necessarily in that order. It brought me material I could never have gotten otherwise. For the depth and the continuity I needed, however, printed books and journals were still my primary sources. The collections at the New York Public Library and the newer Science, Industry, and Business Library, both in Manhattan, provided much useful information.

Mostly, I would like to thank the staff at my own library in Leonia, New Jersey. My local branch is, happily, part of a countywide system, which widens the collection considerably. Special thanks go to Gina Webb-Metz and Teresa Wyman, reference librarians, who managed to draw in a remarkable selection of materials from across the country.

Many colleagues have helped as well. Some were kind enough to read and comment on sections of the manuscript in progress. Among them: William Dunham, Professor of Mathematics at Muhlenberg College, Allentown, Pennsylvania; Daniel Curtin, Professor of Mathematics at Northern Kentucky University, Highland Heights, Kentucky; Professor Joseph W. Dauben, Ph.D. Program in History, City University of New York, New York City; Siegmund Probst, Professor of Mathematics, University of Hannover, Germany; and Dirk van Dalen, Professor of Philosophy, Utrecht University, the Netherlands.

Some of the material was available only in the original French, which I, unfortunately, cannot handle. I also needed help with some German translations and even a few pages in Spanish. Help in these areas came from Daniel Curtin, mentioned previously; J. D.

Nicholson, independent scholar, Baltimore, Maryland; Eric J. Simon, Professor of Psychiatry and Pharmacology at New York University Medical Center, New York City; Fred Stern, independent scholar and lecturer, Leonia, New Jersey; and Laura Mausner, translator, Teaneck, New Jersey.

Yet others helped in some way: by sending me useful material, answering specific questions, or talking through ideas with me. These include Susana Mataix, independent scholar, Madrid, Spain; Stephen Gaukroger, Professor of the History of Philosophy and the History of Science at the University of Sydney, Sydney, Australia; Michael Sean Mahoney, Professor of History at Princeton University, Princeton, New Jersey; Rüdiger Thiele, Professor of Medicine, University of Leipzig, Germany; Richard Bronson, Professor of Mathematics at Fairleigh Dickinson University, Teaneck, New Jersey; Richard Lyons, Professor of Mathematics at Rutgers University, Piscataway, New Jersey; Marshall Hurwitz, Professor Emeritus of Classics, City University of New York, and Arthur Peck, friend, polymath, and retired psychiatrist, Tenafly, New Jersey.

Additional thanks go to my editor, Stephen Power, who believed in the project and provided needed encouragement along the way; to my agent, Faith Hamlin, for her continued and helpful support; and to Fay Klein, who was always there when I needed her.

Introduction

When my editor at John Wiley & Sons suggested that I do a book on great feuds in mathematics, I was not excited by the idea. I had taken some mathematics in school for my master's degree in physics, but that was long ago and I had not used any of it for a very long time. Furthermore, I knew nothing of mathematics' history. Most of all, though, my idea of mathematics was just plain old-fashioned. Mathematics, I felt, is a cold, logical discipline where questions can be decided, if not quickly, at least objectively and decisively. As opposed to, say, politics or religion, or even science, there is little room for human emotions and sensitivities. How could there be feuds in mathematics?

Still, I consulted with an acquaintance, a professor of mathematics, and asked him about the idea. He shook his head and, without giving it a second thought, said, "You'll be lucky if you come up with two feuds."

This fitted in with ideas I could recall from earlier readings. Bertrand Russell, for example, had written, "Mathematics, rightly viewed, possesses not only truth but supreme beauty—a beauty cold and austere, like that of sculpture, without appeal to any part of our weaker nature, without the gorgeous trappings of painting or music, yet sublimely pure, and capable of a stern perfection such as only the greatest art can show."[1]

Isn't it strange how we continue to find what we want to find? As I searched further, I came up with a similar idea from another of my favorite authors, Morris Kline: "Keen minds seeking to establish new systems of thought on the basis of certain cogent knowledge were attracted by the certitude of mathematics, for the truths of mathematics . . . had really never been challenged or been subject to the slightest doubt by the true scholars. Moreover, mathematical demonstrations carried with them a compulsion and an assurance that had not been equaled in science, philosophy, or religion."[2]

I came very close to giving a definite no to my editor, but he, happily, persevered, and so did I. I came across a later book of Kline's, called *Mathematics: The Loss of Certainty* (1980), which told a very different story. I dug some more. The subject, I began to see, allowed for questioning and conflict.

Slowly, as I continued to read, to study, and to talk with others, I began to get the idea that mathematicians, no less than politicians or clerics, are human and, as such, are susceptible to the same emotions, ranging from envy and prejudice to ambition, pride, sibling rivalry, and the irresistible urge to be right. Something of interest was definitely going on here.

As I continued my research, the problem seemed to be not a lack of material but too much. I had to choose among more feuds than I needed for the book. I chose as a beginning point the middle of the 16th century, when a feud between two extraordinary men just jumped out at me.

Their story involves a book—*Ars Magna or The Rules of Algebra*—which has been called one of the great scientific masterpieces of that era. Indeed, it has been credited with giving a jump start to the new sciences of the Renaissance. Contained within the book was a method for solving cubic and quadratic equations. All would have been well except that its author, Girolamo Cardano, was challenged by another Italian, Tartaglia, who claimed not only that credit for one of the basic equations belonged to him, but that Cardano had promised him—as a Christian and a gentleman—that he would not publish until Tartaglia published it first. The row was a splendid one and served as a logical beginning for my new journey.

I knew from my earlier work with feuds that a major source of controversy has to do with questions of priority. Clearly, mathematicians are not in this field for the money, but if they come up with a real advance, they want credit for it. This is true today, and it was no less true in the 17th century. The Newton-Leibniz affair (chapter 2) was such a priority battle. Newton came up with the calculus first but kept it close to his chest. Leibniz published it first, and his method was somewhat easier to use and was put to use first. Who deserved the credit? They battled fiercely, and one of them—using methods that were even then rather underhanded—definitely came out ahead in a personal sense. How their respective countries fared afterward tells a different story.

I was on my way. I was finding all sorts of feuds. Some had to do with pure personal antipathy, a remarkable example being that of the Swiss Bernoulli brothers, two of the world's leading mathematicians (chapter 3). Things started out peacefully enough; in fact, the elder was the younger's teacher. Yet a fierce competition for mathematical supremacy arose between them, and it erupted into public mathematical challenges from one to the other. When a son of one of them developed to the point where he became a threat, he, too, was given similar treatment. But it is also likely that the competition drove these mathematicians to improve their methods, to do better work than they might have done otherwise.

A feud of sorts can also arise because of the very different viewpoints of two individuals. This was the case with J. J. Sylvester, a respected 19th-century British mathematician, and Thomas Henry Huxley, an equally eminent British scientist. Huxley made important contributions in zoology, geology, and anthropology but seemed to have a hole where his mathematics should be. Hence he could argue that "mathematics knows nothing of observation, nothing of experiment, nothing of induction, nothing of causation." It is, in short, "useless for scientific purposes" (chapter 5).

Mathematicians were outraged and felt that Huxley had to be challenged. They chose Sylvester as their champion. The battle between Sylvester and Huxley took place in a rarified space and centered around their two very different points of view. Their discussions and statements would have an effect on the teaching of both science and mathematics in Great Britain and the United States.

All of the feuds so far have been between highly respected, well-placed players. In the case of Georg Cantor, we have a very different kind of battle, one in which there is a clear underdog (chapter 6), but the underdog happens to have been one of the most inventive mathematicians in the field's history. This was both his glory and his difficulty. Cantor had been lucky enough to study with three of Germany's most illustrious mathematicians. He was unlucky, however, in that one of the three was Leopold Kronecker, a well-known but highly conservative professor of mathematics. Cantor's troubles started when he began to move out in several bold directions.

Cantor had in fact opened a wild new world of mathematics. He created set theory, a new concept that turned standard arithmetic on its head. Infinity had long been considered an intangible, incomprehensible puzzle; he not only argued for a real, concrete infinity but even found a way to deal with it mathematically. Yet the bolder his moves, the more they seemed like "mathematical insanity"* to Kronecker, his once-friendly and supportive teacher. The difficulties faced by Cantor in trying to establish his audacious new mathematics, as well as his own reputation, tell a heartrending story worthy of a soap opera.

In the early days of set theory, Cantor had used fairly haphazard methods in determining what elements could go into the creation of sets, leaving it open to attack, and Kronecker was by no means the only one to criticize Cantor's work.

In an attempt to help strengthen set theory, a young German mathematician, Ernst Friedrich Ferdinand Zermelo, came up with a critical item that, according to some mathematicians, saved the day. Officially called the axiom of choice, it also set off a storm of controversy, leading one historian of mathematics to label it the "notorious axiom."†

---

*The term *mathematical insanity* was used by Eric Temple Bell, an important early historian of science. Is it perhaps too strong? We'll find out in chapter 6.
†An axiom in mathematics is a concept or idea that is so obvious we can accept it without having to prove it, and indeed we can even build a logical system upon it.

Among the most vociferous objectors was the Frenchman Émile Borel. The arguments back and forth, by Zermelo and Borel, as well as by their followers, spell out some of the more interesting aspects of the continuing history of set theory (chapter 7).

Still, for a while it had seemed that everything was going to become explainable in terms of set theory, that set theory would become the foundation of all mathematics. In 1901, however, Bertrand Russell—a well-known British philosopher-turned-mathematician—asked a simple question, yet it shook the foundations of set theory and all it stood for in the wider world of mathematics. For it had no answer and was therefore a paradox, or contradiction.

This paradox and others like it had a variety of effects, especially on people interested in the foundations of mathematics, for it began to appear that the whole structure of their beloved discipline was shaky or perhaps was built on a weak foundation. Clearly, the traditional view of mathematics as an exact, logical, and certain discipline had been badly eroded. Starting around the turn of the 20th century, a fairly large group of mathematicians became engaged in studies along this line, but they divided into several mutually antagonistic groups. These gradually formed into three main groups, or schools.

The first school we talk about is logicism, whose main exponent was Bertrand Russell (chapter 8). Russell believed that pure mathematics could be built on a small group of fundamental logical concepts, and that all its propositions could be deduced from a small number of basic logical principles. He also hoped to deal with the paradoxes, which he attempted to do by introducing some new approaches to the problem. Yet Russell had built much of his work on the foundation supplied by Cantor's set theory, and Henri Poincaré—a world-class French mathematician—had been, after Kronecker's death in 1891, the prime opponent of Cantor's mathematics. Result: Poincaré turned his guns on Russell's logicism. Though the two men had high respect for each other, they had no hesitation in attacking the other's position.

The other two schools that arose at about the same time were intuitionism and formalism, whose leaders were L. E. J. Brouwer and David Hilbert. In this battle, all sorts of differences, including the players' nationalities, came into play. When the battle enlarged to bring in supporters, Albert Einstein, who chose to remain neutral, described it as the War of the Frogs and the Mice (chapter 9).

In the final chapter we take a look at a question that has bedeviled and intrigued mathematicians for ages: are mathematical advances inventions or discoveries? Though interesting in its own right, it also leads to an ongoing battle that rages even now about how mathematics should be taught.

Here, then, is a book on great feuds in mathematics. We will see that mathematics is not as objective and certain as it was long thought to be, and that mathematicians are subject to the same frustrations and petty emotions as the rest of us.

Perhaps the difference between what the public sees and what shows so clearly in this book can be explained by an image put forth by Reuben Hersh. Hersh, a professor of mathematics at the University of New Mexico, pictures mathematics as rather like an excellent restaurant. In the front, the dining area, the customers are served clean, well-manicured mathematical dishes; in the back, however, the mathematicians are actually cooking up their new knowledge in a messy, chaotic atmosphere that includes the hot tempers, the disorder, the turmoil, and the failures as well as the successes.[3] We'll be concentrating on the kitchen area, in the back.

1

Tartaglia versus Cardano

Solving Cubic Equations

In 1545, Girolamo Cardano, an Italian physician and mathematician, set the world of mathematics abuzz with a book on algebra. Referred to today as *Ars Magna or The Rules of Algebra*, it is still considered by many scholars to be one of the scientific masterpieces of the Renaissance.

What was so important about an algebra book?

Ars Magna began with some introductory material, including standard solutions to linear and quadratic equations. But then it jumped into uncharted territory and laid out for the first time a complete procedure for solving cubic and biquadratic (third-degree and fourth-degree) algebraic equations.

The book was in truth a stunning achievement and was to play an important role in stimulating the growth of algebra in Europe during

7

most of the remainder of the 16th century. It was not until the arrival
of mathematicians at the level of François Viète (1540–1603) and René
Descartes (1596–1650) that the book's contributions were superseded.

But its impact didn't stop at mathematics, for the Renaissance was
also a formative period in the world of science, and Cardano's book
played a role there as well. As the eminent mathematician and scholar
Morris Kline explains, "Many people credit the rise of modern science
to the introduction of experimentation on a large scale and believe
that mathematics served only occasionally as a handy tool. The true
situation . . . was actually quite the reverse. The Renaissance scientist approached the study of nature as a mathematician. . . . There was
to be little or no assistance from experimentation. He then expected
to deduce new laws from these principles."[1]

By energizing a long-dormant mathematical field, then, Cardano
also provided fuel for the advance of science. The mathematics historian Ronald Calinger, for example, sees Cardano as one of the
architects of the new science of the Renaissance. As a result, the *Ars
Magna* has been compared to Vesalius's *On the Structure of the Human
Body* and to Copernicus's *On the Revolutions of the Heavenly Spheres*, both
of which appeared at about the same time.

Vesalius was a Belgian, however, and Copernicus was Polish.
Surely, Italian mathematicians swelled with pride that one of their
own had made such an extraordinary contribution to the advancement of their discipline.

Certainly, they did—with one major exception. Almost immediately after the publication of *Ars*, an Italian mathematician generally
known by a single name, Tartaglia, began attacking Cardano.
Though Cardano had stated very clearly in the text, and in several
places, that credit for the solution to one of the basic cubic equations
belonged to Tartaglia, this, said Tartaglia, was not the point. Filled
with rage at what he saw as Cardano's treachery, Tartaglia maintained that when he had shown Cardano that solution, Cardano had
promised faithfully—as a Christian and a gentleman—that he would
not reveal it until Tartaglia published it first.

To understand Tartaglia's objections and the strange outcome of
the resulting dispute, we must travel back to the beginning of the
16th century.

Rebirth

The European Renaissance, which actually began during the 14th century, was a rebirth, a reawakening of European mind and culture after a thousand years of sleep. Artists and scholars, especially in Italy, were rediscovering the riches of the past and were adding to them. More slowly, but just as surely, science, technology, and finally mathematics began to awaken as well.

The first stirrings in mathematics began in algebra toward the end of the 15th century. As with other aspects of the Renaissance, these were largely a rediscovery of earlier work, in this case the remarkable achievements of the earlier Greek, Arab, and Hindu mathematicians, who had solved linear and quadratic equations (equations of the form $ax + b = c$ and $ax^2 + bx + c = d$) many centuries earlier.

Arab mathematicians had solved some cubics as long ago as the 9th century and perhaps earlier, but these were geometrical solutions, or even guessed solutions, for specific numerical problems. Still badly needed, and actively sought, was a solution for the general cubic ($ax^3 + bx^2 + cx + d = 0$). Luca Pacioli, the author of the most influential book in Italian mathematics prior to *Ars*, maintained (1494) that such a solution could not be found.

Then, sometime between 1510 and 1515, Scipione del Ferro (1465–1526), a professor of mathematics at Bologna, came up with the first algebraic solution to a cubic equation. He had developed an algebraic formula for solving the "depressed cubic," a specific third-degree equation that lacks its second-degree term. In other words, he had come up with a general solution for $x^3 + ax = b$, with a and b positive. It was a real breakthrough, but he kept it a virtual secret for a dozen years and maybe more! What could explain such surprising behavior?

First, the turn of the 16th century was not a time of "publish or perish." There were no peer-reviewed journals; there was no Internet. In fact, the more likely scenario was for the discoverer of a new solution to keep it close to his chest and to use it publicly, if he did at all, only when it could somehow prove advantageous.

For example, the idea of tenure lay long in the future, so academic appointments in mathematics could be tenuous. Chairs were held by virtue of eminence and reputation, and public challenges might come at any time. Contests sometimes resulted in public disputations that could be large, contentious affairs, often attended by the disputants' students and supporters. In some cases, the contests attracted large crowds and even passionate betting. Del Ferro apparently believed that if challenged with a list of problems to be solved, he could always use his method as a powerful counterpunch.

History does not tell us whether del Ferro ever used the solution in this way, but we do know that upon his death in 1526, his papers, with the solution, passed on to his son-in-law and successor, Annibale della Nave, and, more important, to one of his pupils, Antonio Maria Fior.[2]

Fior felt he was now in possession of a valuable treasure, and he returned to his native city of Venice with the objective of establishing himself as a teacher of mathematics. He let it be known that he had a special ability regarding cubic equations. Yet he kept hearing that maybe it wasn't so special, that someone else had this ability, too. The name he heard was that of Tartaglia, a teacher of mathematics in Venice and Verona who was making a name for himself in public debates and who had also made some claims regarding cubic equations.

Fior thought about issuing a public challenge to Tartaglia. If Tartaglia's claims were exaggerated, which was quite possible, it would be a good way to build up his own reputation while tearing down that of this pretender.

Tartaglia

The likelihood that Tartaglia would later have anything to offer mathematics had seemed small when he was born in Brescia, in northern Italy, in 1499. His father was a mailman, and the family was poor. Whatever Tartaglia learned of mathematics and science, he picked up on his own.

He was not always called Tartaglia; he was christened Niccolò Fontana. But this was a dangerous time, and about dozen years after his birth the town was sacked by the French. Young Niccolò took a slashing wound to the mouth and the palate and came very close to death. Although his mother brought him through with careful, tender care, the wound caused permanent damage to his speaking apparatus, with the result that he was nicknamed Tartaglia, meaning the Stutterer. The name stuck.

Tartaglia eventually settled in Venice and made his living as a teacher of mathematics. As with other such teachers, he did the best he could to keep his name before the public by participating in public contests and disputes, and he seems to have had some success in these contests. The 19th-century biographer Henry Morley wrote of Tartaglia that he may "fairly enough be said to have become wholly by his own exertions a distinguished mathematician, as it is also certain that he grew up to be like many other self-taught men, rugged and vain."[3]

Tartaglia had implied to a colleague that he could solve a numerical equation of the form $x^3 + cx^2 = d$. This was enough to act as a direct challenge to Fior. Early in 1535, Fior challenged Tartaglia to a public contest. They came to an agreement: each would propose 30 problems to the other. Whoever should solve the most problems after 30 days would be the winner. There was little fear of collusion, for there was nowhere else to turn for help.

No one knows how much Tartaglia really knew at the start of the contest, but by the evening of February 13, 1535, he was able to solve both types of numerical cubic (with and without the x^2 term), a tremendous accomplishment. This meant he could solve all 30 of Fior's problems. On the other hand, Tartaglia had apparently been aware that Fior was capable of solving only the depressed cubic, and he had designed his questions around this form. Fior therefore had little success with Tartaglia's questions.

Tartaglia was the clear winner, and little was heard of Fior after that.

Tartaglia's fame grew, and the numbers of his students grew apace. Again, with our modern mindset, we would expect Tartaglia to have published his newfound, or newly developed, technique for solving the cubic, but no, he, too, kept it close to his chest. Morley would later complain, "His new rules concerning cubic equations he maintained as his private property, cherishing them as magic arms which secured to him a constant victory in algebraic tilts, and caused him to be famed and feared. . . . That," he continued, "was a selfish use to make of scientific acquisitions, with which no scholar of the present day [1854] would sympathize, and which, also, in the sixteenth century, would have been thought illiberal . . . even by our erratic and excitable Cardan."[4]

As we shall see, he might better have said, "*especially* by our erratic and excitable Cardan."

Yet mathematics, remember, can be a useful tool, as well as a fascinating puzzle, and with many of the Western powers fighting for control of Italy, the application of mathematics to ballistics was a hot topic. Tartaglia applied himself to it. So it was that in 1537, Tartaglia's eye was not on cubic equations but on ballistics, and he published a successful book on the subject; his *Nova Scientia* described both new methods and new instruments.

His star, clearly, was rising, but it was to take a most unexpected turn. For just as Fior was falling out of the ring, Girolamo Cardano, a far more dangerous opponent, was stepping in.

Cardano, Renaissance Man

Though Girolamo Cardano was close in age to Tartaglia, his early life had been very different. Born in Pavia in 1501, he later recalled, "My father, in my earliest childhood, taught me the rudiments of arithmetic. . . . After I was twelve years old he taught me the first six books of Euclid, but in such a manner that he expended no effort on such parts as I was able to understand by myself."[5] Cardano's father was a well-educated lawyer, a lecturer in geometry, and a friend of Leonardo da Vinci, who was himself interested in mathematics.

Although Cardano's father wanted him to study law, Cardano, despite showing clear ability in mathematics, leaned toward medicine as a career. He began his university training at age 19 at the University of Pavia. By age 21, he was debating in public and lecturing on Euclid. After transferring to the University of Padua, he received his medical degree there at age 25.

Øystein Ore, one of his biographers, writes that "he was quick tempered and vindictive and often unable to control his anger. At times this involved him in brawls of the most serious kind."[6] So we should not be surprised when Cardano tells us in his autobiography that from the years 1524 through 1547, he was engaged almost constantly in lawsuits—and, he claims, won them all. Perhaps he should have gone into law?

Apparently not. Smart and personable, he was at the same time building a strong reputation in medicine and, by the late 1530s, had become possibly the most sought-after physician in northern Italy. It seems, however, that medicine at the time presented different career choices than it does today. In 1537, he was invited to teach medicine at Pavia, but he refused, "for there seemed little hope of receiving pay for the work."[7] His medical income came basically from private patients and from patrons.

Cardano was a most unusual character, though, and the term "Renaissance man" could have been invented to describe him. For in addition to medicine, he made his mark in several other, quite different, fields. He was, for example, an inveterate gambler, and he published a very popular and useful handbook of gambling, which included some advanced work on probability, as well as detailed information on cheating.

He also cast horoscopes for the rich and powerful. This was a common and widespread practice among people skilled in mathematics and astronomy, and although Cardano was proud of his capability, it got him into various kinds of trouble. In one case he cast a horoscope for Edward VI, the boy king of England, that was just plain wrong, and later he did one for Jesus, which turned out to be a very bad idea.

Cardano was born into a superstitious family and carried on the tradition. At the same time, he made some good observations in medicine and in natural history (what we know today as science). As Cardano himself put it in his autobiography: "If then you place the number of important branches of learning at thirty-six, from . . . any acquaintance with [twenty-six of] them I have refrained." He then modestly admitted, "To ten I have devoted myself."[8]

We are concerned with his mathematical work, and specifically with his *Ars Magna*. Certainly, he brought the algebraic solutions for cubic and biquadratic equations into the open, which was a tremendous accomplishment. How much of this did he owe to Tartaglia and others? Let's take a closer look at how the book came into being.

Preliminaries

After hearing of Tartaglia's success over Fior, Cardano had asked Tartaglia for permission to publish his solution to the cubic equation in his (Cardano's) own forthcoming book on mathematics, promising to give full credit to Tartaglia.

Tartaglia's initial answer was that he was planning to write a book himself in which he would spell out the rule. When? He couldn't say, for he was occupied with other things at the moment, including, first, his ballistics work, and then a translation of Euclid. Not being easily dissuaded, and convinced of the importance of the solution, Cardano kept after Tartaglia, using a variety of entreaties.

A series of letters has come down to us that alternates between sharp and friendly. Initially, for example, Cardano labeled Tartaglia as greedy and unwilling to help mankind.[9] Then Cardano sharply criticized some work in Tartaglia's book on ballistics. Tartaglia fired back an answer that included "But, in believing that you can demonstrate miraculously by your ridiculous opposition that I am wrong, you have only demonstrated, I will not say, that you are a great ignoramus, but that you are a person of poor judgment."[10]

Cardano shifted gears: "You should not imagine that my sharp words were caused by enmity. . . . I really wrote that abuse to excite you to a reply."[11]

As part of his campaign he invited Tartaglia over for a friendly visit to his home, figuring, correctly, that he would have greater leverage that way. Cardano proclaimed that he was interested in the solution for purely academic reasons. What finally did the trick was his use of an important name, that of his patron Alphonso d'Avalos, Marchese del Vasto. D'Avalos—the Spanish governor of Lombardy, whose capital is Milan, and commander of the imperial army stationed in the area—was one of the most powerful men in Italy. In a letter to Tartaglia, Cardano wrote, "I must in the first place state that

I have held you in good esteem, and as soon as your book on Artillery appeared, I bought two copies, of which I gave one to Signor the Marchese."[12]

In another letter, dated March 13, 1539, Cardano wrote that his excellency "commanded me at once to write the present letter to you with great urgency in his name, to advise you that on receipt of the same you should come to Milan without fail, for he desires to speak with you." Tartaglia, well aware that friendship with d'Avalos could be very useful, finally assented: "I go thither unwillingly: however, I will go."[13]

As it turned out, d'Avalos was not in Milan when Tartaglia arrived. Deliberate deceit? Or just the result of the busy schedule of an important man? It's hard to say. Øystein Ore, in his biography of Cardano, evaluates the former possibility and points out that this "would have been a very complicated and dangerous scheme. The ruse could readily have backfired on Cardano if Tartaglia on the strength of the invitation had written directly to d'Avalos."[14]

Nevertheless, Cardano did manage to pry Tartaglia's secret out of him, but Tartaglia was not so foolish as to just hand over the solution. What he gave Cardano was his "rule" for solving the depressed cubic but not his "demonstration," which would be the general method, or in modern terms the proof that the rule produced the solution. In addition, he gave it in the form of a cryptic verse, though he may later have clarified it for Cardano.

In May 1539, Cardano's *Practica Arithmeticae Generalis* appeared—without Tartaglia's solution. It contained some errors, which Tartaglia was happy to point out. In fact, he made fun of both Cardano and the book, which Cardano revised in future editions. Then Tartaglia began hearing rumors about a new book on algebra. Cardano denied the rumors, and things were quiet for a while, but he was indeed working on such a book.

Cardano was in fact a prolific writer. By the end of his life, he had published thousands of pages in various disciplines. *Ars* was to have been volume 10 in an encyclopedia of mathematics—which he never completed and of which not much remains. *Ars Magna* is a shortened version of his original title (*Artis Magnae Sive de Regulis Algebraicis*). In English, it means *The Great Art*, to distinguish it from other, more elementary works, such as his own earlier one on arithmetic. He was also well aware that the solution to the cubic equation would be of great importance to its success. So, along with his very able assistant, Ludovico (also spelled Lodovico) Ferrari, he put in several years puzzling out the meaning of the verse and expanding the implications when he began to understand it. For, as we'll see, the *Ars* presentation was no simple restatement of Tartaglia's rule.

Ars Magna

The material on cubic equations first appears in chapter 11, which is titled "On the Cube and First Power Equal to the Number." This is interesting on several counts. The rule Tartaglia gave Cardano covered the three basic forms of the depressed cubic. In modern terms, these would be: $x^3 + bx = c$, $x^3 = bx + c$, and $x^3 + c = bx$. The three forms were necessary because mathematicians at the time did not use negative coefficients, and this precluded use of the single, general form $x^3 + ax + b = 0$. In addition, our modern algebraic notation still lay in the future, and most of the mathematical statements were verbal. The chapter title, for example, refers to the specific form that we would today write as $x^3 + bx = c$.

Cardano's book also employs considerable geometric material. In fact, as William Dunham puts it in his fine book *Journey through Genius*, "His argument was purely geometrical, involving literal cubes and their volumes. Actually, the surprise here is minimized when we recall the primitive state of algebraic symbolism and exalted position of Greek geometry among Renaissance mathematicians."[15]

In each chapter, then, Cardano first gives a geometrical demonstration of a specific numerical cubic equation, then a verbal rule for solving that general type of equation, then one or more sample problems and solutions using the rule. Because the use of zero and negative coefficients still lay in the future, Cardano is forced into spelling out 13 different cubic equations, all with positive coefficients, and with a separate chapter for each type.

Furthermore, these geometric solutions are by their nature both roundabout and somewhat cumbersome, and because the notation at the time was primitive, the book makes for difficult reading today,

and we needn't go through any of his demonstrations. Yet it's worth showing how his rule works for a specific example of the depressed cubic, which he gives in chapter 11.

In the book, Cardano first presents a general statement of the rule for each chapter, which would work for any numerical example of this type; then he gives a specific example and shows it at work in that example. I'll combine them. I'll state his rule and, to save space, will simply insert the results of this particular example in square brackets as we go.

In modern notation, the example is $x^3 + 6x = 20$ and his rule, in translation, begins:

"Cube one-third the coefficient of x $[\{1/3(6)\}^3 = 2^3 = 8\}]$; add to it the square of one-half the constant of the equation $[10^2 = 100]$; and take the square root of the whole $[\sqrt{108}]$. You will duplicate this, and to one of the two you add one half the number you have already squared and from the other you subtract one-half the same. You will then have a *binomium* $[\sqrt{108} + 10]$ and its *apotome* $[\sqrt{108} - 10]$. Subtract [the cube root of the] apotome from that of the binomium and you will have the value of x:

$$\sqrt[3]{\{\sqrt{(108)} + 10\}} - \sqrt[3]{\{\sqrt{(108)} - 10\}}."$$

Cardano doesn't bother to spell out the answer, but the mathematicians among you will realize that the solution to this complicated expression is nothing more than the number 2.

Not all of his examples ended with whole number answers. In some examples, he found himself with imaginary roots. Though he was baffled by them, he did acknowledge their existence.

A Sacred Promise?

There is no question that Cardano's contribution to the field was considerable. The question is, just how perfidious was his treatment of Tartaglia? The answer remains as elusive as ever. First, Ore points out that none of Cardano's contemporaries expressed their displeasure at the time, even though the details of the affair were widely discussed. The negative points of view seem to have arisen later, in the 18th and the beginning of the 19th centuries.[16]

Was there a sacred promise of secrecy? Many, perhaps most, writers say yes, but this is mainly based on Tartaglia's claim. In the year following publication of *Ars*, Tartaglia published his own work *Quesiti et Inventioni Diverse* (*New Problems and Inventions*), which included what he maintained were word-by-word accounts of their meetings.[17] Many writers depend on this publication and quote the following promise that Tartaglia says Cardano made to him: "I swear to you by the sacred Gospel, and on the faith of a gentleman, not only never to publish your discoveries, if you will tell them to me, but also I promise and pledge my faith as a true Christian to put them down in cipher, so that after my death nobody shall be able to understand them."[18]

Others point to the fact that Ludovico Ferrari, Cardano's secretary and assistant, was also present when Cardano and Tartaglia met, and that Ferrari later swore, just as vociferously, that Cardano never made such a promise. In fact, Ferrari claimed that in general, Tartaglia's accounts of the earlier proceedings, made in the heat of his rage, were doctored.

Alan Wykes, a modern biographer, goes even further. He argues that Cardano had figured out the algebraic equations by himself, or at least without help from Tartaglia. In *Ars*, says Wykes, "by a slip of pen or memory, he [Cardano] wrote that Tartaglia had communicated the discovery to him and given him permission to use it." But, Wykes argues, "It may perhaps not have been a slip but a muddled attempt at a generous gesture on Cardano's part."[19]

Yet between the time of Tartaglia's visit to Cardano and the appearance of *Ars*, six years had elapsed. During that period, Cardano and Ferrari, having heard rumors of the existence elsewhere of such a solution, had traveled to Bologna in 1543 and visited their colleague Annibale della Nave. There they were given permission to examine the papers of Scipione del Ferro, and they learned that del Ferro and not Tartaglia had been the first to solve such an equation algebraically. In such case, they reasoned, even if Cardano had been sworn to secrecy, that promise was no longer valid.

Even before publication of *Ars*, there are suggestions that something of this sort was afoot. For example, after one of Tartaglia's refusals, Cardano wrote again with another request but added, "I

should like to save you from the illusion that you are the first man in the world . . . , I want to write to you amiably to dissolve the fantasy that you are so great. I will lovingly let you know even through your own words that in knowledge you are rather in the valley than near the summit of the mountain."[20]

Was he here suggesting that Tartaglia had not been the original possessor of the solution? Why not come right out and say so? This was 400 years ago, and things may have been done differently.

In any case, Cardano was careful not to claim credit for discovering the rule. In three places, he included citations for earlier work on the cubic equation. Near the beginning of chapter 1, for example, he writes, " In our own days Scipione del Ferro of Bologna has solved the case of the cube and first power equal to a constant, a very elegant and admirable accomplishment. Since this art surpasses all human subtlety and the perspicuity of men's minds, whoever applies himself to it will believe that there is nothing that he cannot understand. In emulation of him, my friend Niccolò Tartaglia of Brescia, wanting not to be outdone, solved the same case when he got into a contest with his [Scipione's] pupil, Antonio Maria Fior, and, moved by my many entreaties, gave it to me."[21]

Cardano repeats almost the same words at the beginning of chapter 11: "He [Tartaglia] gave it to me in response to my entreaties," but adds that Tartaglia withheld the demonstration: "Armed with this assistance [the rule], I sought out its demonstration in [various] forms. This was very difficult. My version of it follows."

And his version is indeed different, and much expanded, from what Tartaglia had given him—or from whatever he had gotten from any other source. More specifically, with the help of his secretary/assistant, Ludovico Ferrari, he uses the solution he started with (for the depressed cubic, obtained from Tartaglia or whoever) as a stepping-stone. By employing appropriate substitutions, which reduced them to the known case, he had found solutions to the three additional cubic equations. Then, again with the help of Ferrari, to whom he gives appropriate credit, Cardano also shows how these cubic solutions could be used as a foundation on which to build solutions for biquadratics, or equations of the fourth degree. In addition, Cardano points out that cubic equations should have three roots.

A Battle Not-So-Royal

Despite Cardano's acknowledgments, Tartaglia was in a rage. The following year he published his book *New Problems and Inventions*, mentioned earlier. The first half does indeed contain solutions to problems that had been put to him over the years, but the latter part is devoted entirely to a full-out attack on Cardano and *Ars*. It includes reproductions of their correspondence, along with his comments. It is a bitter, and powerful, attack. He publicly heaps scorn on Cardano's mathematical abilities. Wykes describes it as a "swingeing attack . . . denying ever having given Cardano permission and accusing him of theft."[22]

As Tartaglia thought would happen, he got a response, but not quite what he was expecting. For it seems we are dealing here with not just two, but with three quite disputatious types. Ore writes, "The Renaissance abounds in impulsive and hotheaded geniuses and Ferrari ran true to form. He had such a temper that even Cardano at times was afraid to speak to him, and one day when he was seventeen years old he came home from a brawl missing the fingers on his right hand."[23]

On February 10, 1547, Ferrari, rather than Cardano, responded with a printed cartello, a challenge to Tartaglia to meet him in a dispute on almost any scientific topic, maintaining that Tartaglia had "written things which falsely and unworthily slander the above-mentioned Signor Gerolamo [*sic*] (Cardano), compared to whom you are hardly worth mentioning."[24] Ferrari sent the cartello to a variety of scholars and dignitaries all across Italy, so that Tartaglia could hardly refuse. Ferrari's attack was strong; he argued that Tartaglia had built up his reputation by attacking others; that, ironically, he had published a proof in his book that was stolen and for which he had not given credit; and that the book was full of errors.

There was an interchange. Tartaglia complained that he wanted to meet Cardano, not Ferrari, the pupil. Ferrari, however, acting on Cardano's behalf, insulted Tartaglia strongly enough that Tartaglia had little choice but to acquiesce. Ferrari also made the point that Cardano attributed the solution to del Ferro and Fior, both of whom knew it before Tartaglia, and that there was no oath of secrecy.

In any case, a public contest did take place the following August. The details are vague, but Tartaglia seems to have withdrawn after a brief but possibly vituperative battle, and Ferrari was declared the victor. Cardano didn't even attend.

Though few details of the actual contest have come down to us, the apparent outcome is supported by the later results. Tartaglia lost a promised position in his hometown of Brescia, whereas Ferrari received a variety of good offers: he was invited to lecture in Venice, Tartaglia's home territory, and went on to become a professor of mathematics in Bologna. Tartaglia now had additional fuel for his bitterness.

It seems odd that Cardano, so proud of his learning and having done some solid work in the mathematics field, was apparently content to have Ferrari act as his champion. Ore points out, "Cardano began his university disputes in his student days, apparently with much success. . . . Cardano must have been well equipped for the debates; he had a quick wit, a good memory, and a sharp tongue. According to his own account he became so proficient in these mental duels that his opponents had little chance of victory, or even defeat with honor."[25] Ore is referring mainly to medical contests; did Cardano think less of his mathematical abilities?

As had happened much earlier with Fior, Tartaglia lost some public esteem and some reputation. This, however, only ramped up his bitterness, which was aimed directly at Cardano, and he retreated into a corner of the ring—fuming, waiting, watching, and plotting.

A First Attempt

What Tartaglia saw, though, was an opponent whose star continued to rise, and who was therefore not an easy target. Wykes writes, "The name of Doctor Cardano rang through the halls of philosophy, astrology, mathematics, science and medicine. Books, tracts, and treatises by the score came from the press of Petrieus [his publisher]."[26] How this must have galled Tartaglia.

He first tried to entrap Cardano via a complicated scheme that involved his [Cardano's] friendship with Gonazaga, the governor of

Milan and an opponent of the pope, and the old pope's addiction to astrology. Tartaglia had an idea that he could make trouble by engineering an offer to Cardano to enter the service of the pope as an astrologer and a physician. Perhaps he thought he could generate some sort of charge of political intrigue.

Cardano was riding high, however, and in no need of patronage. He turned down the offer for personal reasons. Tartaglia then tried to implant the idea that Cardano had intended to offend His Holiness by his refusal. Tartaglia also produced a copy of Cardano's earlier horoscope of Jesus in an attempt to convince the pope that it was blasphemous. The situation was briefly troubling to Cardano, but this first attempt at revenge by Tartaglia petered out early in the 1550s. He retreated and began plotting another campaign.

Cardano's star, however, was still shining bright. He held the chair of medicine at Pavia, which provided material success, time to write, and considerable prestige. Offers of all kinds arrived constantly. Tartaglia held himself in check until, eventually, his patience was rewarded. Fate stepped in and dealt him a full hand.

Cardano was blessed in many ways, but not with his children. Three of them, two sons and a daughter, were to pave the way to his undoing. Thanks to Cardano's eminence, his daughter Chiara had made a good marriage to Bartolomeo Sacco, who came from a noble lineage. Unfortunately, Chiara was a promiscuous young woman, and Sacco not only sought an annulment but wanted recompense from Cardano for this "worthless baggage."[27] During the years 1557 and 1558, Cardano found himself enmeshed in both legal and church battles that began to tear at his once-illustrious name.

Though he managed to continue his work as a doctor and a philosopher/scholar/writer, he was deeply enmeshed in his daughter's problems when trouble came from another direction. In December 1557, his elder and favorite son, Giambattista, married. Cardano would have expected the boy to marry well, but he chose the daughter of a down-and-out family, which Cardano was expected to support and for which he ended up supplying most of the funds.

The details are not clear, but it must have been a marriage made in hell, for two years later, Giambattista's wife was dead of arsenic poisoning, and Cardano's son was arrested for her murder. Cardano did all he could to prove his son's innocence, but Giambattista confessed and was executed. Cardano never got over this, and it may well have affected his mind, for he started to become suspicious of attempts on his life.

It is, of course, possible that some of these were actual attempts to destroy him in one way or another. Envy of the famous is not uncommon. Add to the pot his cantankerous character, however, and it's easy to believe that there were some people who did what they could to pull him down.

Now add to his real and imagined difficulties new problems with his other son, Aldo. Aldo turned out to be a drinker and a gambler. He lived with Cardano for a while, but the two did not get along. Early in the 1560s, Aldo moved out, but, having gotten deeply into debt, he broke into Cardano's house and was caught. Cardano wanted nothing further to do with him. Here was yet another disaster with his offspring.

Again he lost himself in his writing, but by now attempts were being made to eject him from his chair of medicine at Pavia, and then at the even more prestigious college in Milan. In 1563, his name was removed from the list of scholars qualified to lecture, and he was accused of various crimes. He was actually exiled from Milan, which included Pavia, and he left there at the end of 1563 in the depths of despair, his fortune gone, his books impounded.

A Shadowy Hand

The rest of the decade saw no improvement for Cardano. His country, dominated by Spain and decimated by wars, labored under heavy taxation. The Spanish Inquisition, well underway, was a potent force. Scholars of all sorts were under suspicion, but somehow Tartaglia had managed to place himself satisfactorily. Cardano could find no employment, and, according to Wykes, "it was Tartaglia who was the instigator of most of the refusals that met him in College and University. It was simple enough, with the network of the Inquisition flourishing in city, vineyard, village and public square, to keep a shadowy hand on the shoulder of any citizen, great or small."[28]

This was just the warmup, though. On October 13, 1570, almost a quarter of a century after publication of *Ars Magna*, Tartaglia served up a double blow. Using Cardano's own son Aldo as an informant as to Cardano's whereabouts, Tartaglia handed him to the Inquisition. Tartaglia had been collecting evidence against Cardano for years. Among this "evidence" was Cardano's rejection of the pope's invitation that he become the pope's astrologer and physician. Tartaglia pointed to the "sarcasm" evident in Cardano's comment that "His Holiness by his study of astrology has surely raised himself among the greatest of such scientists and has no need of help from such as myself."[29]

Cardano's horoscope of the life of Jesus was also damning, as were a variety of other statements that, taken out of context, could be construed as blasphemous. In one of his publications, for example, he had suggested that God is a universal spirit whose benevolence is not restricted to holders of the Christian faith. Today he might be admired for such an ecumenical statement; at the time, it was apparently a dangerous idea.

And so it went. Cardano, fortunately, was not subjected to torture or put to death, but he was thrown into jail. He sought desperately for help and was able to reach out to an official in the church, Archbishop Hamilton, who had in the past asked to be called upon if need be. The archbishop came through for Cardano, who was released a few months later. It was just in time, for not long after, the archbishop's own fortunes changed; he was captured by the forces of Mary, Queen of Scots, and beheaded.

Tartaglia finally had had his revenge. Cardano lived on in obscurity in Rome, where he worked on his autobiography, which is one of the works that has come down to us in full. He probably never knew, and just as well, that his daughter Chiara had died of syphilis, and that it was Aldo who had betrayed him to the Inquisition and who was rewarded with an appointment as official torturer and executioner in Bologna.

Cardano died on September 20, 1576. Less than a year later, Tartaglia followed him to the grave.

Who Won?

Ore, who has studied Cardano carefully, argues, "His originality in other fields has sometimes been questioned, but *De Ludo Aleae*, his obscure and somewhat disreputable book on how to win at cards and in dice games, contains indisputable proofs of his genius."[30] In addition, Wilhelm Gottfried Leibniz (see chapter 3 in this book) maintains, "Cardano was a great man with all his faults; without them he would have been incomparable."[31]

As for Tartaglia, Ore believes that "had Tartaglia never existed, the science of mathematics would not have been deprived of a single great or fertile idea."[32] When Tartaglia died, an attempt was made to assemble and publish his unpublished papers. Oddly, none could be found that even mentioned the solution to the cubic equation.

Nevertheless, says Ore, Tartaglia was no slouch. Under other circumstances, his star might well have shown brighter. As Ore puts it: "His great tragedy was the head-on collision with the only two opponents in the world [Cardano and Ferrari] who could be ranked above him."[33]

While both Tartaglia and Cardano paid dearly in this battle, there is no question that mathematics came out a winner.

2

Descartes versus Fermat

Analytic Geometry and Optics

Pulling together work he had quietly labored over for decades, the French philosopher and mathematician René Descartes finally published in June 1637 the book that was to make him famous. His *Discourse on Method* is considered by modern historians a major landmark in several areas, including philosophy, the history of science, and, especially, mathematical thought. David E. Smith and Marcia L. Latham, who did the definitive translation of the mathematical portion of the book, compare it to Newton's *Principia* and argue that it contributed to the great renaissance of mathematics in the 17th century.[1]

Thanks largely to this work, Descartes is commonly given credit for unifying algebra and geometry and even for the creation of analytic geometry; indeed, the Cartesian coordinate system is named after him. That's what we remember today. This being mathematics, one might expect that with Descartes' introduction of this new material, the resulting discussion was precise, well-defined, and devoid of emotional fireworks.

That, however, was not to be. What should have been a straightforward mathematical discussion turned into a mind-boggling combination of Greek tragedy and modern spy novel, with a plotline that included the Catholic Church, several of the top mathematicians of the time, and the vastly differing personalities of Descartes and his main opponent, Pierre de Fermat. The result: a drawn-out battle with a clear winner and loser, but with the ironic outcome that the winner learned little or nothing from the battle, while the loser was inspired by the battle to come up with a major principle in science and to provide important grounding for the development of the calculus.

The mathematics historian Michael Sean Mahoney writes, "Few scientific debates in history reveal so much of the personalities of the participants, or the extent to which personal factors can influence rational discourse."[2]

Descartes

Born in 1596, Descartes received his education in the early years of the 17th century. It was an exciting time, the age of Galileo, Kepler, Harvey, Gilbert, and Francis Bacon; of Shakespeare and Montaigne. It saw the beginnings of what later came to be called the Age of Reason and spawned people like Newton and Leibniz, Milton and Moliere. Yet the standard education was still largely based on the classical curriculum.

So it was for Descartes. His first five years of schooling were devoted almost entirely to Latin, Greek, and classical literature. At the age of 10, he entered La Flèche, a prestigious, highly disciplined Jesuit school, where he spent the next eight years. Among the main subjects were the works of Aristotle, but mainly as viewed through the glasses of the Jesuit fathers. It was basically via these secondhand readings that Descartes had his initial grounding in philosophy.

He was an apt pupil, but when he left—at age 18, in the year

1614—he appears to have come away with "the discovery at every turn of my own ignorance"[3] and utter disdain for the philosophies then being taught. "Philosophy," he wrote later in the *Discourse*, "affords the means of discoursing with an appearance of truth on all matters, and commands the admiration of the more simple."[4] In 1616, he earned a law degree at the University of Poitiers, ~~to which~~ ~~but~~ he subsequently paid little attention~~.~~ ~~the~~ Law .

Certainty and Method

Though contemptuous of the existing philosophies, his interest in the subject had been piqued. He began to wonder: how do we know what we know? How can we be sure that what we know, or think we know, is correct? How can it be that so many have studied so much, and yet so much of what we hear and learn is so wrong or so uncertain?

He was convinced, for example, that Copernicus was correct—that although the sun *appears* to be circling the earth, the earth actually circles the sun. In that case, however, can we depend on our senses for learning anything in the world around us?

Then he learned that Copernicus's book *De Revolutionibus*, in which Copernicus had put forth his heliocentric theory more than half a century earlier, had been censored and suspended by the Catholic Church until it would be corrected, or at least formulated so that the idea was put forth as a hypothetical one. For although Copernicus's idea was a good one, it went against church teachings, and there was no real, physical proof of its correctness. If there had been concrete proof of the concept, the church could not object. Descartes began to develop a passion for certainty, which was later to become central to his entire corpus of work.

Several ideas began to gel. Of all the subjects he had studied, mathematics seemed to provide the one real road to certainty. For, he believed, mathematics depends totally on rational thinking; it protects against errors introduced by the senses or even by measurement and experiment.

In 1618, he met and began working with Isaac Beeckman, a

teacher and educational administrator who, like Descartes, had a deep interest in the association between mathematics and the physical world. Under Beeckman's guidance and encouragement, he began to focus intensely on mathematical and mechanical problems. He spent some time in the army and appears to have done some mathematical work on military architecture while there. As a kind of gentleman-soldier, he had free time to spend on his studies; still, he was not happy in the army. He complained of being idle and of being in the company of uneducated people. He left the service early in 1619.

He was highly respectful of the mathematics of the ancient Greek mathematicians, such as Pappus and Diophantus. He also, however, suspected that they deliberately held back in their presentations; that is, that they showed solutions to certain problems but kept their methods secret, in much the same way as the algebraists of Cardano's day had done.

In 1619, he wrote, "When I attended to the matter more closely, I came to see that the exclusive concern of mathematics is with questions of order or measure, and that it does not matter whether the measure in question involves numbers, shapes, stars, sounds, or any other object whatsoever. This made me realize that there must be a general science that explains everything that can be raised concerning order and measure irrespective of subject matter, and that this science should be called *mathesis universalis*—a venerable term with a well-established meaning—for it covers everything that entitles these other sciences to be called branches of mathematics. How superior it is to these subordinate sciences both in usefulness and simplicity is clear from the fact that it covers all they deal with. . . . Up to now, I have devoted all my energies to this *mathesis universalis* so that I might be able to tackle the more advanced sciences in due course."[5]

By 1619, then, he already knew that he was destined to create a philosophical system that would utilize deductive procedures as rigorous as Aristotle's, but based on his own thinking and development. He had, as he put it, discovered the foundations of a marvelous science (*mirabilis scientiae fundamenta*), on which could be based a complete philosophical system that would provide a road to learning and study, paved with certainty and clarity.

Descartes spent the next two decades developing these ideas and expanded his coverage to include nothing less than the whole world. By 1628, in fact, he had begun working on *Le Monde (The World)*, his broad mechanical explanation for the way much of the world works. In 1633, however, when he had the manuscript ready, he heard that Galileo had been hauled up before the Inquisition for espousing Copernicus's heliocentric theory. Some of Descartes' ideas in his own book, he feared, would displease the church as well. Good Catholic that he was—apparently his training at La Flèche had had some effect—he chose not to publish rather than rock the ecclesiastical boat. He felt, too, that he would rather not weaken his statements. If he was to successfully refute ancient authors such as Aristotle, his work had to be presented with at least as much certainty as their systems.

Right along, though, he was developing both his philosophy and his explanations for how things work in the world. His vortex theory postulated a matter-filled universe in which the motion of any body can be caused only by contact with another. It had enormous influence—for a while. Its advantage was that it provided a mechanical explanation for many heretofore puzzling phenomena, as well as for explanations that previously had depended on spirits and ghosts.

In the interim, he had produced the makings of several books, but, for various reasons, not one was published. Until the *Discourse*, in fact, he had no publications. His fear of offending the church was one reason. Another was that the scientific journal still lay in the future. In France, at least, its place was taken by Father Marin Mersenne, a scholarly priest whose cell in Paris had become a meeting room for some of the foremost French mathematicians of the day, including Blaise Pascal, Pierre Gassendi, Gilles Personne de Roberval, and Jean Beaugrand. Beaugrand would play a major role in the initiation of the coming conflict with Fermat.

Mersenne corresponded with other eminent mathematicians and was instrumental in facilitating the interchange of mathematical developments throughout Europe. He was often referred to as France's "walking scientific journal," and it was by his efforts that Galileo's work became known outside his own country.

Mersenne and Descartes became friendly in 1622, and Mersenne began to spread the word that a new and promising philosopher/

mathematician was developing. By 1626, thanks largely to the communications efforts of Mersenne, Descartes' reputation had grown substantially, even though he had not yet published a single word.

Discourse on Method

Descartes' *Discourse on Method* (1637) was actually a pastiche of works he had composed at various stages of his studies, though it also included some new material. Its full title was *Discourse on the Method of Rightly Conducting One's Reason and Seeking the Truth in the Sciences*. The introductory part, often referred to as the Discourse, contained the basic idea and the philosophical rationale for the entire book.

In this opening section, he laid out four laws that served as the guiding principles of his approach. "The first of these," he wrote, "was to accept nothing as true which I did not clearly recognise to be so: that is to say, carefully to avoid precipitation and prejudice in judgments, and to accept in them nothing more than what was presented to my mind so clearly and distinctly that I could have no occasion to doubt it."[6] His philosophy, then, was one of systematic doubt. Of one thing he could be certain, however, and so it was that his famous statement was born: "I think, therefore I am." One of his biographers, Stephen Gaukroger, explains, "Descartes begins by showing that, provided one's doubt is sufficiently radical, there is nothing that cannot be doubted, except that one is doubting, and this requires that there be something which exists that is doing the doubting."[7]

Of course, doubt was not enough. I noted earlier that he felt he could use mathematics as a foundation on which to build. As he put it: "I was especially delighted with the Mathematics, on account of the certitude and evidence of their reasonings: but I had not as yet a precise knowledge of their true use; and thinking that they but contributed to the advancement of the mechanical arts, I was astonished that foundations, so strong and solid, should have had no loftier superstructure reared on them."[8]

His book was not a full description of the world, but it did suggest that all physical phenomena can be explained mechanically, which turned out to be a very potent concept. Following the relatively brief introduction were three essays showing some examples of how his method could be used toward this end. Two would lie at the center of his dispute with Fermat.

The first, "Dioptrics," deals with the nature and the properties of light. Descartes saw light not as motion but as a pressure or a "tendency to motion" that was transmitted instantaneously (or very close to it) through a kind of elastic medium. This came logically out of his vortex theory: he believed that we experience the light ray in very much the same manner as an impression of movement or resistance that would travel instantly from the point of action to a blind man's hand through his stick.

Therefore, he felt, light will travel instantaneously, or nearly so, through optical media, and also that its speed will actually be greater in a denser medium such as water than in air.

He also thought of both reflection and refraction in terms of material collisions; he assumed that in reflection the light is thrown back like an elastic ball from an elastic surface, and that similar reasoning holds for refraction, except that in this case the ball breaks through the surface.

By this reasoning he came up with his law of refraction, which stated that the ratio between the sine of the angle of incidence of a light ray and the sine of its angle of refraction is a constant:

$$\sin i/\sin r = n^9.$$

Descartes had worked the law out mathematically, an impressive accomplishment.[10]

The second essay, "Meteors," was perhaps the first real attempt at a scientific work on the weather. It included a description of how rainbows are produced, which he based on his law of refraction.

Descartes' "Geometry"

In the third essay, "Geometry," Descartes had put together what turned out to be his main legacy in mathematics. He presented, and solved, one of the most difficult problems bequeathed to the mathematical world by the ancients. It had been thought up by the Greek geometer Apollonius in the 3rd century B.C. His contemporary Euclid and then Pappus some six centuries later did more work on it. Yet in spite of much effort by them and many later mathematicians, no one before Descartes had been able to solve it completely—that is, to provide a general solution.

Using his method, Descartes had attacked the problem some years earlier and had solved it in a matter of weeks. The problem, as stated by Descartes, was: "Having three, four or more lines given in position, it is first required to find a point from which as many other lines may be drawn, each making a given angle with one of the given lines. . . . Then, since there is always an infinite number of different points satisfying these requirements, it is also required to discover and trace the curve containing all such points."[11]

J. L. Coolidge restates the problem: "If from a point in a plane, line segments be drawn to meet four given lines of that plane at pre-assigned angles, and if the product of the first and third segments bear a constant ratio to the product of the second and fourth, then the locus of the point in question is a conic."[12]

Descartes' main contribution was to treat the problem algebraically and generally. The example he gave involves four lines, but his method could be generalized to n lines, and it could be reduced to one in which all we need to know are the lengths of certain lines. These lines are the coordinate axes, the lengths of which provide the abcissas and ordinates of needed points.   *abscissas ?*

His association of equations and curves is one essential feature. Also, he located his points and curves on a single coordinate system, a step that had not been taken before. It was not, however, a rectangular coordinate system as we know it today. He used only a single unmoving axis with a moving ordinate, which was not necessarily vertical, but this was a major step nonetheless.

In essence, he had found a way to apply the algebra of Cardano and those who came after him to the geometry of the ancient mathematicians. As Descartes put it in "Geometry," "Here I beg you to observe in passing that the considerations that forced ancient writers to use arithmetical terms in geometry, thus making it impossible for them to proceed beyond a certain point where they could see clearly the relations between the two subjects, caused much obscurity and embarrassment, in their attempts at explanation."[13]

Using one of his own rules, he recognized the necessity of getting rid of the many numbers and "incomprehensible [geometric] figures" that overwhelm the procedure, as it was done before his work. Recall that equations in Cardano's time were still laid out in verbal terms. Toward the end of the 16th century, François Viète, a French lawyer attached to the court of Henry IV, had made some important advances in algebraic notation and in the general improvement of the theory of equations. Viète (1540–1603) was among the first to represent numbers by letters and to introduce at least the beginnings of a general symbolism, but his algebra still differed from ours in an important way. He still saw problems in a geometric sense; he saw the product of two line segments, for example, xx, as an area. As a result, there was even a question as to whether equations of degrees higher than three made any sense at all.

Descartes began his "Geometry" thus: "Any problem in geometry can easily be reduced to such terms that a knowledge of the lengths of certain lines is sufficient for its construction."[14] He saw the product of two lines, say a and b, not only as a rectangle but also as a line. Similarly, terms like x^2 and x^3 could be seen as line segments and not as a square and a cube. Result: he was able to restate a geometric problem in algebraic terms and solve it algebraically.

He also made a significant contribution to the theory of equations. He wrote, "If, then, we wish to solve any problem, we first suppose the solution already effected, and give names to all the lines that seem needful for its construction—to those that are unknown as well as to those that are known. Then, making no distinction between known and unknown lines, we must unravel the difficulty in any way that shows most naturally the relations between these lines, until we find it possible to express a single quantity in two ways."[15] (That is, to solve the resulting simultaneous equations.)

Thus, if two curves were considered in the same system of coordinates, their points of intersection could be gotten by solving the equations of the two curves and finding the roots common to their two equations.

So now we can express relations in terms of only two variables. For example, as Emily R. Grosholz states it (as in the Pappus problem), "Distances between the fixed lines and a point C on the locus can be expressed in the form $ax + by + c$, and the condition which determines the locus can be expressed as an equation in two unknown quantities. For three or four fixed lines, this equation would be a quadratic equation, for five or six lines, a cubic, and so forth, the introduction of every two lines making the equation one degree higher."[16]

Descartes also brought the symbolism very close to what we use today, whereby lowercase letters at the end of the alphabet represent unknown factors, and letters at the beginning are used for constants and known terms.

Finally—and largely as a result of his "Geometry"—he is commonly credited with the creation of analytic geometry—that is, a geometry in which a point is a set of numbers located in what is now known as a (Cartesian) coordinate system, and a geometric construction can be thought of as a collection of points and described by equations or formulas. It took time for this to happen, however. In fact, the subject was not even given the name *analytic geometry* until the 19th century.

In a letter to Mersenne in 1637, he said, modestly, "I do not enjoy speaking in praise of myself, but since few people can understand my geometry, and since you wish me to give you an opinion of it, I think it would be well to say that it is all I could hope for, and that in *La Dioptrique* and *Les Meteores*, I have only tried to persuade people that my method is better than the ordinary one. I have proved this in my geometry, for in the beginning I have solved a question which, according to Pappus, could not be solved by any of the ancient geometers."[17]

As I stated earlier, he was wrong here. The ancients had solved the problem, but only for one or two specific cases. What he did was produce a general solution, which they had not done. All in all, Descartes felt that he had produced a solid, useful—and unique!—document. As far as he was concerned, all this material was new to the world. He fully believed, for example, that he had developed, for the first time, a specific method for finding the truth—that is, knowledge that is both solid and certain. Though the method would work for various kinds of knowledge, he was particularly interested in finding the "truth" in the sciences.

Dismay—and Conflict

We can only imagine, then, the shock, dismay, chagrin—and even anger—he must have felt upon seeing criticisms of his book by some of the major mathematicians of the day. Among these comments were some by a barely known lawyer and amateur mathematician named Pierre de Fermat. In order to better understand what happened when he received Fermat's comments, though, we must back up a bit.

An independent spirit like Descartes was not one to mince his words. In 1636, a contemporary of his, Jean Beaugrand, had published a book called *Geostatics*, and Descartes had issued a brutal criticism of it. Was Descartes' action at least in part payback for some earlier criticisms of Descartes by Beaugrand? Perhaps. In any case, the publication of Descartes' *Discourse* gave Beaugrand an opportunity to avenge himself. In the spring of 1637, he managed to obtain an advanced copy of Descartes' "Dioptrics" and began a vicious campaign against it. He circulated the manuscript among his own colleagues, including Fermat, apparently in hopes that it would receive some serious criticism even before it was published.

Fermat, in all innocence, and having no idea that the manuscript had been obtained unethically, issued what he thought was a purely scientific critique. He had several objections. As a firm believer in the importance of experiment, he objected to Descartes' reliance on mathematics to study physical phenomena. Specifically, he objected to Descartes' investigating the "inclination to motion" by his mathematical examination of the motion of a tangible physical object (the ball against the elastic sheet).

Another problem, he said, was in Descartes' demonstration and proof of his law of refraction; Fermat argued that it was in fact no proof at all. He stated that Descartes' result was implicit in the assumptions, that "of all the ways of resolving the determination [that is, the tendency] to motion, the author has selected only the one that leads to his conclusion; he has thereby accommodated his means to his end, and we know as little about the subject as we did before."[18]

Fermat had written his comments in the form of a letter to Mersenne. He had started it with a suggestion that we mathematicians can often "find what we are seeking by groping about in the shadows." He then amplified this with his own objections, as noted previously, and he closed with an offer. "We must seek the truth in common," he wrote, and he, Fermat, would be happy to help Descartes in his search.[19] It's not hard to imagine Descartes' reaction when he read this.

What Fermat apparently did not know was that Descartes, upon hearing of Galileo's run-in with the Inquisition, had pulled back his publication of *Le Monde*, in which he had spelled out his physical theories far more explicitly than he could in the *Discourse*. Therefore, as Mahoney explains, "Descartes' *Dioptrics* appeared without the cosmological treatise on which it was based. The short account of the nature of light that opens the *Dioptrics* could not replace the more extended and more carefully argued theory of *The World*, or *On Light*, which Descartes had withdrawn from publication. . . . Moreover, the crucial steps in his derivations of the laws of reflection and refraction depended on the laws of motion presented in the suppressed treatise. Without the precise context of *The World*, the appearance of those laws in the *Dioptrics* seemed arbitrary at best; they were not set out properly until the publication of the [Descartes'] *Principles of Philosophy* in 1644. . . . Only against that background can one understand Fermat's critique, for it focused precisely on the points that required the fuller context."[20]

At first, then, Descartes was not overly concerned. Fermat, he figured, had simply not understood what he was getting at. Nor did Descartes have any idea of what kind of competition he was facing. By the end of 1637, however, two exchanges took place that changed things considerably. Fermat had seen Descartes' "Geometry" and had expressed surprise at the lack of any work by Descartes on maxima and minima, which he felt was so important that it should have been included in a work on mathematics. Fermat had thereby sent to Mersenne his own work on this area, which included methods of finding maxima, minima, tangents to curves, and, very important in the conflict, his own work on analytic geometry. Descartes saw this work just prior to publication of his *Discourse*. Though Fermat had come at the Pappus problem from another approach, and even though this was not his latest work, his methods and procedures were uncomfortably close to Descartes' own.

By this time, other comments and criticisms were coming in. To all of these, Descartes responded mainly with anger and contempt. William R. Shea has described his response nicely.[21] I summarize it here: The French mathematicians who criticized his "Geometry" were dismissed as "two or three flies";[22] Roberval was described as "less than a rational animal";[23] Pierre Petit as "a little dog";[24] and Hobbes as "extremely contemptible."[25] Jean de Beaugrand's letters were only good to be used as "toilet paper."[26]

Descartes' reaction to the comments by Fermat was similar. In a letter to Mersenne, he compared Fermat to Ennius, an earlier Roman poet, and himself to Virgil. Virgil had been quoted in Donatus's *Life of Virgil* as feeling that he was gathering gold out of Ennius's shit.[27]

In general, as the mathematics historian J. F. Scott puts it, "Descartes firmly believed that he had nothing to learn from his contemporaries in any branch of mathematical knowledge, and in particular he leaves his readers in no doubt that he did not rate the achievements of Fermat very highly. In a letter to Mersenne he declared that none of his critics . . . had been able to achieve anything of which the ancient geometers were ignorant." Referring to these critics, he specifically mentioned "M. vostre Conseiller De Maximis et Minimis," meaning, of course, Fermat.[28]

Fermat, the Hesitant Amateur

Born in 1601, five years after Descartes, Fermat was the son of a prosperous leather merchant who also served as second consul of his town. His mother, too, had a high social standing. After a solid secondary education, he received his law degree from the University of Toulouse in 1631. Trained as a classical scholar, he was fluent in Latin and Greek and became interested in "restoring" the lost works of ancient scholars. Among them were the mathematical works of two great Greek mathematicians, Apollonius and Pappus. Still, to this point there was no indication that he would turn out to be one of the great mathematicians of his time.

Named a judge in Toulouse in 1638, he went on to become the king's councilor in 1648. Though he spent most of his life in Toulouse, he had lived some years in Bordeaux. It was during his time in Bordeaux, at the age of about 20, that he became fascinated by the work of Viète.

By the mid-1630s, he got to know Mersenne and was invited to correspond with the Paris group. By the spring of 1636, he had already been working on the ideas that would eventually cause so much heartache for Descartes.

Diffident in his writing, he tried to have it both ways. He wanted recognition, yet did not want to open himself up to criticism. Roberval offered to edit and publish some of Fermat's work, but Fermat refused categorically. Still, as he was by now in steady correspondence with his fellow mathematicians, he was becoming better known. Among these mathematicians was Beaugrand, who in fact prided himself in having "discovered" Fermat.

It was not surprising, then, that Beaugrand had sent a copy of "Dioptrics" to Fermat and asked for his comments. These would, naturally, be funneled through Mersenne. When Mersenne received them, he saw trouble brewing and dithered for several months before transmitting them to Descartes, in spite of Descartes' words in the *Discourse* that asked for comments. Finally, however, Mersenne bit the bullet and sent them, with the results we have already seen.

Attack and Response

By the end of the year, then, Descartes was dealing with a lot. He even began to suspect that he was the subject of a concerted plot to destroy his brainchild. Referring to some of his opponents, such as Pascal and Roberval, he wrote to Mersenne, "I beg you to see if they have not erased the words: *E jusques a* [*E until a*] and replaced them with *B pris en* [*B includes en*]. Because this is the way they cite me in their writing, in order to corrupt the sense of what I said."[29] Fermat's response had been even more troubling.

Descartes, who did not take well to criticism in any case, was faced with a devastating set of events. His life's work had been criticized, to his mind severely. Mahoney writes that Descartes' mathematics, as put forth in both "Dioptrics" and "Geometry," "enjoyed his most jealous protection, for the new method of mathematics had been the source of the larger philosophical method of the *Discourse*. To attack it, to correct it, or to find something not already in it, was to impugn Descartes' whole program. To claim to have achieved similar results independently of, or earlier than, Descartes was to question the uniqueness of [Descartes'] mission."[30] Yet by the end of the year, Fermat had done exactly that. Furthermore, in Descartes' eyes, these objections were coming from a disciple of his hated rival Jean Beaugrand.

Descartes was not one to calmly accept this state of affairs. A contentious man under any circumstances, he felt that this was a situation that needed correction.

Descartes took a closer look at Fermat's work and in January of 1638 began to retaliate. Whether he had in mind a specific mission—namely, to destroy Fermat's growing reputation—or was merely reacting to the criticisms is hard to say. He advanced some specific objections to Fermat's mathematics—for example, to his work on determining the tangent to a cycloid—but then, in a letter to Mersenne, he charged Fermat with deficiencies both as a mathematician and as a thinker. Fermat's methods were defective, he said, and thereby had little value. He went further, suggesting that Fermat was indebted to him for much of what he had developed.

This was particularly unfair. Although Fermat's analytical geometry came to its final form around 1635, it is well known that he had developed various aspects of it much earlier. In addition, by 1635 he had already applied his method to the locus problem that served as the starting point for both Fermat's and Descartes' work in this area. The general consensus is that Fermat was totally unaware of Descartes' work at that time.

Others by this time were entering the lists: Roberval and Pascal had sided with Fermat; Claude Mydorge and Girard Desargues took Descartes' side.

Though Fermat is now recognized as one of the great mathematicians of his era, his tendency to skimp on detail in his writings probably made it easier for Descartes' charges to stick. Even early on, before his trouble with Descartes, some of Fermat's peers were irritated by what looked to them like a supercilious attitude. He would, for example, throw out challenges that he said he had solved, but he gave no details. Had he really solved them? The activity in our own day, centering around Fermat's last theorem, is a case in point. Fermat had written in the margin of a book that he had solved this enormously difficult problem, but he added that he had no room to spell out any details. Only now, almost three centuries later, can anyone claim to have finally come up with a final proof.[31]

Anger Builds

Fermat was to take what appeared at first to be a much less contentious stance than Descartes. In December 1637, he wrote to Mersenne, "First of all I assure you that it is not due to envy or rivalry that I continue this little dispute but only to find the truth; for which I think M. Descartes will not believe me of ill will, the most so since I am well aware of his outstanding ability. . . . Before I enter into the discussion, I will add that I do not wish my letters to be more widely shared than with those with whom an intimate conversation is possible; this I entrust to you."[32]

In February, he wrote again, "I understand from your letter that my reply to M. Descartes was not appreciated, in fact that he decided to comment on my method of maxima and minima and on tangents, in which nevertheless he will find Mssr. Pascal and Roberval of the opposite opinion. Of these two things [Descartes' objections], the first [re refraction] does not surprise me, for matters of physics can always raise doubts and lead to disagreement. But I am astonished by the latter [Descartes' denigration of his methods] since it is a truth of geometry and I maintain that my methods are as certain as the first proposition in the [Euclid] *Elements*. Perhaps being presented plain and without proof they were not understood or else they seemed too easy for M. Descartes, who has made so many paths and followed such a difficult road to tangents in his *Geometrie*." [This is actually a snide remark, as will become clear later.]

"I will send you nothing more for M. Descartes since he puts such strict rules on an innocent exchange. I will be content to tell you I have found no-one here who does not agree with me that his 'Dioptrique' is not proven."[33]

And again: "I await, if you please, the reply M. Descartes made to the difficulties I showed you with his *Dioptrique* and his remarks on my 'Treatise on maxima and minima and on tangents.' If there is some rancor, as he seems to fear there is . . . , that should in no way keep you from showing them to me, for I assure you they will have no effect on my spirit, which is far away from vanity, so that M. Descartes cannot rate me so low that I would not rate myself lower. It is not that my accommodating nature obliges me to retract a truth that I already know, but I want to let you know my mood. Oblige me, please, by not hesitating to send me his writings, about which I promise you in advance to make no reply."

In the same letter, later on: "Whenever you wish my little war with M. Descartes to end, I will not be grieved, and if you arrange for me the honor of his acquaintance, I would be greatly obliged to you."[34]

In Descartes' "Geometry," he had also presented a general method for finding the normal to a curve at any point, and he was very proud of it. Unfortunately for him, Fermat's method was far more direct and closer to modern treatments. Except for simple algebraic curves, what could take pages of complex computation by Descartes could be accomplished far more expeditiously by Fermat.

Descartes began to see some of this as he gave it some thought. By then, he was also less sure of a "conspiracy" than he had been earlier. In mid-June 1638, he wrote to Mersenne, using the flamboyant terms common in those days: "I beg him [Fermat] most humbly to excuse me and to consider that I did not know him. Rather, his *De maximis* came to me in the form of a written challenge on the part of him who had already tried to refute my *Dioptrics* even before it was published, as if to smother it before its birth, having had a copy of it that had not been sent to France for that purpose. Hence it seems to me that I could not have replied to him in words any softer than I used without evincing some sort of laxity or weakness."[35]

That was what he wrote to Mersenne. Yet he had too much at stake, and his bitterness was now fed by his searing hatred of Fermat's friend and defender Gilles Personne de Roberval, so he did not really ease up. Among his more famous comments was one that he made to a colleague, Frans van Schooten, who related it later in a letter to Huygens in 1658. M. Fermat, he said, is a "Gascon." The word can be translated in several ways: it could refer to the region Fermat was from, but it is more likely to mean troublemaker or, most likely, braggart. "I am not [a Gascon]," Descartes continued. "It is true that he [Fermat] has found many pretty, special things, and that he is a man of great mind. But, as for me, I have always endeavored to examine things quite generally, in order to be able to deduce rules that also have application elsewhere."[36]

Among his other charges was that Fermat's method of finding maxima and minima and his rule of tangents were not the result of strict a priori deduction. More important, he argued that Fermat's reputation was built largely on a couple of lucky guesses. This, applied to one of the great mathematicians of the day, was particularly galling to Fermat and his followers. Unhappily, Descartes' reputation gave strength to the rumor, and by the early 1640s, Fermat was seen by some of his peers as having operated by trial and error, rather than by careful and logical analysis.

Things May Not Be What They Seem

To this point, we have to agree with E. T. Bell, who wrote of their mathematical disagreements: "It seems but natural that the somewhat touchy Descartes should have rowed with the imperturbable 'Gascon' Fermat. The soldier was frequently irritable and acid in his controversy over Fermat's method of tangents; the equable jurist was always unaffectedly courteous."[37]

This in fact seems to be a common reading of Fermat's character. In W. W. Rouse Ball's classic *A Short Account of the History of Mathematics* (1908), for example, we find: "The dispute was chiefly due to the obscurity of Descartes, but the tact and courtesy of Fermat brought it to a friendly conclusion."[38] In addition, Mahoney describes Fermat as "gentle, retiring, even shy. . . . There is much to suggest that he simply did not like controversy and that he shied away from it whenever possible."[39]

In a recent article, however, Klaus Barner, a German professor of mathematics, takes Mahoney to task, as follows: "A stereotype that goes back to Mahoney . . . and has been adopted by more recent authors, is that Fermat was a mediocre conseiller and judge who tried to avoid all social, political and religious conflicts. Nothing is further from the truth. Fermat . . . was an outstanding practitioner who . . . stood up for justice and humanity without shrinking from confrontations with the mighty."[40] Mahoney, and perhaps others, had been misled by an incomplete reading of a 1663 report, and that report in turn had relied in part on a wicked untruth perpetrated in Fermat's day by one of his enemies.[41]

It seems likely, then, that Fermat did not shrink from conflict with Descartes, but, as we'll see, his weapons and methods may have been rather more subtle than Descartes' had been.

Continuing Provocation

After the interchanges of the 1630s, the public discord between Fermat and Descartes more or less died down. In fact, it lay quietly for almost two decades. During that time, however, Descartes' attacks on Fermat's reputation were having their desired effect, and Fermat's contributions were increasingly ignored.

Interestingly, while Descartes' reputation continued to grow, it was more in the area of philosophy than in mathematics, for his *Geometry*, by now in book form, was not an easy work to deal with. Ironically, there was some of Fermat's (mathematical) reticence in it. For example, it did not provide full proofs of his work—in order, he wrote, to give others the pleasure of discovering the proofs for themselves. Yet Descartes' colleague, the Leiden mathematician Franz van Schooten, saw the gold in the book. He translated the work into Latin and added considerable explanatory commentary. The revised book came out in four subsequent editions over the years 1649–1695 and had a significant influence on a new generation of mathematicians.

At the same time, however, Descartes' acid pen alienated some of the major mathematicians of his day, including Roberval and Pascal, and he found himself defending himself against charges, for example, from Beaugrand and the British mathematician John Wallis, that he had plagiarized Viète and/or the British mathematician Thomas Harriot (1560–1621). Subsequent studies suggest this was not the case,[42] but the charges had some effect at the time. Descartes was even accused of having used Fermat's work. The most likely scenario, however, is that both men worked independently.

Too, as Mahoney puts it: "A bit more homework on his part prior to publication might have toned down his claim to unprecedented novelty and originality."[43] This might have muted the criticisms.

During the two-decade hiatus in their battle, Descartes continued his work in philosophy and metaphysics and published several well-received works. At the same time, the revised editions of his *Geometry* helped cement his reputation in the mathematical world. He died in 1650, honored and celebrated.

Retaliation

Fermat had not forgotten his hurt, however, and toward the end of the 1650s, he finally had an opportunity to retaliate. Claude Clerselier, an ardent supporter of Descartes, was in the process of preparing an edition of Descartes' letters and asked Fermat for copies of letters he wished to include in his collection. He had copies of the two key letters that Fermat had written to Mersenne for transmittal to Descartes (May and December 1637), but Clerselier had reason to believe there were others. He asked Fermat for copies of such letters. Fermat either misunderstood the request or felt he now had an opportunity to put in a good word for himself. In March of 1658, a little more than 20 years after the original controversy, he composed a long letter to Clerselier, in which he repeated his earlier criticisms and added others.

Mahoney writes, "The restatement was not entirely accurate, as Clerselier, in possession of the original letters, knew full well. Hence, it appeared to him that, by adducing new arguments against

Descartes' derivations, Fermat was seeking to reopen the dispute. He [Clerselier] felt his suspicions confirmed when, in response to his and Jacques Rohault's defense of Descartes, Fermat continued his attack all the more stoutly. The result was a series of letters between Clerselier and Fermat that continued over the next four years."[44] In all, eight letters were exchanged.

And what letters they were. We are especially interested in Fermat's, and I will give only a sample from one of his, for they are difficult to parse. All the letters were written with, as Mahoney calls it, "Baroque politesse," which "barely masks the anger and indignation in which they were written."[45]

J. D. Nicholson, who has translated and studied these letters, adds, "To understand Fermat's letters, you must understand the term *politesse*, a term which in French means right-thinking attention to manners, but in English is likely to mean the use of such manners for less than noble purposes." While Descartes' letters were likely to be direct and unadorned in their criticism of Fermat, the politesse of Fermat was so subtle that the portion of Fermat's letter that follows could easily be taken as praise if not looked at carefully.

For, says Nicholson, while Fermat is saying one thing, he often means something quite different. I will first give the polite translation,[46] and then, with Nicholson's help, I will give just a few of the possible alternate meanings. Too, the Baroque mode of expression was as prolix and excessive in writing as it was in architecture and furniture design. So take a deep breath before you venture into these two paragraphs. Fermat wrote:

> The conclusions that can be taken from the fundamental proposition of M. Descartes' *Dioptrique* are so beautiful and ought naturally to produce such lovely results throughout every part of the study of refraction that one would wish—not only for the glory of our deceased friend, but more for the augmentation and embellishment of the sciences—that this proposition were genuine and legitimately demonstrated, and all the more as it is from these [conclusions] that one is able to say that *multa sunt falsa probabiliora veris* (often, falsehoods are more acceptable than truth). . . .

I begin from there, Monsieur, in order to let you know that I would be delighted if the differences that I have formerly had on this subject with M. Descartes were ended to his advantage. I would have been satisfied in all ways: the glory of a friend whom I have infinitely esteemed and who has passed with good reason for one of the great men of his time; the establishment of a physical truth of the greatest importance; and the easy execution of these marvelous effects. All this seems incomparably more valuable than winning my case; likewise, I would count for nothing the [Latin] phrase "He will win the fight with me," of which the friends of M. Descartes are always able to reasonably comfort his adversaries. I put myself therefore, Monsieur, in the posture of a man who wants to be vanquished. I say it loudly: I bow at last to your superior powers.[47]

In the first paragraph, did Fermat really mean the "glory of our deceased friend" or did he actually have in mind Descartes' vanity or pride? Descartes himself defined glory as a type of "love that one has for one's self."

In the second paragraph, Fermat wrote, "I would be delighted if the differences that I have formerly had with M. Descartes were ended to his advantage." Rather than "I would be delighted," however, he could also have meant "I would feel raped" or "ravaged." Did Fermat really mean "one of the great men of his time," or was he implying "one of the fat pretentious ones"?

There are also several interesting literary references. Consider, in the first paragraph, "Often, falsehoods are more certain than truth." Fermat could simply be saying that Descartes' false scientific idea was more acceptable than Fermat's own correct one, but it could also be a reference to a famous court case argued by Cicero in 81 B.C. A group of men has stolen the inheritance due a rightful heir. To keep the inheritance, they drag the heir into court on a trumped-up charge of murdering his father. Fermat, in other words, was "tried" by many of Descartes' friends in the court of public opinion and perhaps is the heir denied his due.

In the second paragraph, Fermat uses the Latin phrase "He will win the fight with me." Here he's referring to the famous classical conflict between the heroic Ajax and the smooth-talking Ulysses. In the end, Ulysses' silver tongue wins him the spoils, and Ajax plunges his sword into his own chest. It's not hard to guess who is Ulysses and who Ajax.

Another reference is seen in Fermat's line about bowing. The words "I bow at last to your superior powers" is Charles E. Bennett's translation of the first line of the Latin poet Horace's Epode 17, "A Mock Recantation."[48] In this poem, the protagonist is sentenced to spread the fame of another for all eternity. Fermat apparently is not hopeful that his letters will have the desired effect, but this doesn't mean that he isn't going to try to get the credit he believes he deserves.

There is more like this before Fermat actually gets to the corrections he wishes to score with, but it is enough to give the flavor of the letters. Also, note the construction. It's not easy to tease out the real meaning, but the key words are "one would wish . . . that this proposition were genuine and legitimately demonstrated." His contention, in spite of all the fine words, is that Descartes' proposition was not legitimately, or satisfactorily, proved.

Finally, concentrating his attack on Descartes' derivations, Fermat used stronger and more direct language than he had in 1637. For example, he once again attacked Descartes' proof of the sine law, but, says Harvard professor emeritus A. I. Sabra, "not any longer because it is not conclusive, as he believed twenty years earlier, but simply because it now appears to him to be founded on an assumption that is 'neither an axiom, nor is . . . legitimately deduced from any primary truth.'"[49]

In the course of his reasoning, however, Fermat came up with an important idea, which has come to be called Fermat's principle of least time. In essence, it states that nature follows the shortest path, or the least time possible, for a process in nature. Based on it, and on his own, somewhat different assumptions from Descartes', he in 1661 mathematically derived his own sine law. In essence, he stated that in refraction, it was the optical distance—the products of the distances the light traveled and the corresponding refractive indices—that is a minimum.

Ironically, although some of his objections to Descartes' work were valid, his sine law turned out, most writers feel, to agree exactly with Descartes': $\sin i = n \sin r$.

Sabra, however, argues that the sine law did *not* match exactly. He writes that although both laws assert the constancy of the ratio of the sines, Fermat thought that Descartes, in the course of his reasoning, had used the construction $n = v_i/v_r$ ($= \sin i/\sin r$), whereas it was actually $n = v_r/v_i$.[50]

It didn't matter, though. The first reasonable determination of the speed of light was not to take place until 1657, so experimental verification of Fermat's theory was not yet possible anyway. Clerselier argued that it was ridiculous to believe that nature could change its mind as light travels from one medium to the next.

Fermat's approach had led him to deduce that light speed is finite, and that light travels faster in air than in water. Both of these conclusions were brilliant insights and just the opposite of Descartes' results. Science eventually came down on Fermat's side. His principle—later expanded to include maxima as well as minima—is now thought of as a basic law of optics, but none of this could be known at the time.

Sabra concludes, "In the face of this unexpected result [identical laws of refraction], he [Fermat] was willing to abandon the battlefield, as he said, leaving to Descartes the glory of having first made the discovery of an important truth, and himself being content to have provided the first real demonstration of it. With this conditional declaration of peace in his last letter to Clerselier of 21 May 1662, the discussion came to an end."[51]

Comparison

In the two decades between the two episodes, Descartes had pretty much steered clear of mathematics. Fermat, on the other hand, had been busy indeed, which makes Descartes' denigration of Fermat's mathematical abilities both sad and strange. By the 1630s, Fermat had already shown his colors, though not widely and not publicly. That Descartes didn't like his colors was Descartes' problem, not

Fermat's. For in the intervening years, Fermat continued his mathematical efforts and came up with several major advances. First, his principle of least time provided a good foundation for several areas in physics. In fact, the whole of geometrical optics could be based on a properly modified form of his principle.

Fermat also made major advances in number theory and probability and provided groundwork for the calculus, which was to follow soon (see chapter 3 in this book). His work, in fact, strongly influenced a number of later mathematicians, including Jean Bernoulli (see chapter 4). Thanks to his reluctance to publish, however, his own thesis on analytic geometry was not published until 1679, which was 14 years after his death.[52]

So he had some small recognition, but basically it all came after he died in 1665. By 1662, perhaps discouraged that there seemed to be little interest in his new work, he had essentially retired from the field altogether. His activity in number theory during the last 15 years of his life found no resonance among his colleagues, and when he died, there was little of the honor he deserved. Part of the reason was that his feelings about publication had apparently not changed. He wrote to a friend, "I much prefer to know the truth with certainty, rather than to take more time in debates and superfluous and useless contentions."[53]

On the other hand, during these last years the new editions of Descartes' *Geometry* were being published, and, with the clarification and the simplification, his treatise gained new adherents and influence. His science, too, was in ascendance. Ironically, however, it was both men's work in analytical geometry that eventually led to the development of the calculus later in the century, and this in turn led to the decline of Cartesian science.

There is general agreement that Fermat's basic approach to analytic geometry is significantly closer to our own than Descartes' was. Yet to the end, Fermat championed Viète's cumbersome notation, while Descartes' symbolic notation is quite modern.

Both men made major contributions to the advance of mathematics in the 17th century. It is sad that the process generated so much heartache and bitterness.

3

Newton versus Leibniz

Credit for the Calculus

At the turn of the 18th century, Isaac Newton, an Englishman, and Wilhelm Gottfried Leibniz, a German, engaged in a ferocious battle. Having never met personally, they obviously used neither fists nor knives, yet the science historian Daniel Boorstin has christened their dispute "The spectacle of the century."[1] Ernst Cassirer, writing in the prestigious *Philosophical Review*, called it "one of the most important phenomena in the history of modern thought."[2]

It's usually described as a priority battle over the invention of the calculus. With neither money nor a woman at its heart, this hardly sounds like the stuff of a knock-down-drag-out fight, yet it went on for years, becoming more bitter as it proceeded.

What made it such a big deal? First, it engaged two of the greatest geniuses who ever lived. The more familiar genius is Isaac

Newton, who hardly needs any introduction here. Less familiar is Wilhelm Gottfried Leibniz, a German philosopher/mathematician who did important early work on symbolic logic and on the calculus, and in a variety of other fields as well, including especially cosmology and geology.

Second, though few of their contemporaries could understand or follow the workings of the calculus at first, it shortly became clear that this was a new, useful, and general method for dealing with a wide variety of scientific and mathematical problems that had been unsolvable until then.

The feud also had several curious results. For example, it played an important part in the development of the modern scientific paper—specifically, one that is refereed and that includes explicit, clear references to what has been accomplished previously.

Furthermore, it raged on for centuries, fed mainly by the jingoistic behavior of the two men's followers. It even played a part in an ongoing tug-of-war between British and Hanoverian leaders for the throne of England. Among the Hanoverian claims to distinction was, said Leibniz's followers, his invention of the calculus. Newton's supporters derided such claims. One Briton, John Keill, considered these assertions to be attempts at stealing the fruits of Newton's genius.

It was a battle in which Newton used some heavy-duty fighting tactics—some people would use stronger language—and emerged the clear winner. The result was a gray cloud over Leibniz's later years, but, had he lived long enough, he would have seen another, totally unexpected, outcome. Although Leibniz lost the battle, it's probably fair to say that he actually won the war—though you'd hardly know it from the reputations of the two men as they are known today.

Newton

Nothing as complex and far-reaching as the calculus emerges full-blown from the minds of even men like Newton and Leibniz. Fermat's method of finding maxima and minima was already a direct step along the road to differentiation, one of the important routines in the calculus. For a mind like Newton's, however, no one can say for sure how he built the foundation for his work on this area of mathematics. We do know that he read widely in the mathematical literature of the time, and that one of the books he read, digested, and worked through carefully was Descartes' *Geometry*. He also studied Euclid, whose geometry he is said to have found trifling.

Among other authors there was James Gregory, an accomplished Scottish mathematician; Galileo; and Newton's immediate master at school, Isaac Barrow. We have a specific clue: Newton did tell us that he was led to his first discoveries in the field by his reading of *Arithmetica Infinitorum* (1655), a text by the distinguished British mathematician, cryptographer, and cleric John Wallis that dealt with the quadrature of curves (finding the area beneath curves).

Newton began his mathematical work while a student at Trinity College, Cambridge, and he earned a bachelor's degree in June 1665. Then an attack of plague shut down the university for 18 months. He simply continued his studies, on his own, at his family home in Woolsthorpe, a small town some 30 miles southeast of Nottingham in central England. He may, however, have made a brief visit to the university at some point during this time, perhaps to do some reading and/or experiments.

His enforced home study time was apparently the best thing that could have happened, at least for Newton. During this period, spanning the years 1665 to 1666, he built the foundations for his work in optics, celestial mechanics, and mathematics, including the foundations of the calculus. As part of this work, he extended Wallis's work on the use of infinite series. Newton realized, and capitalized on, the fact that many mathematical functions can be expressed as infinite series. Using them, he was able to generate general expressions for the lengths and the tangents of curves, along with a method for handling quadrature problems (calculating the areas bounded by curved lines). Practitioners of the calculus will recognize here the beginnings of their craft.[3]

At this point, Newton, an unknown youth in his mid-20s, had already sailed past his own teacher at Cambridge, and even past Wallis, one of the top mathematicians of the day. Mathematicians until then had thought of the path of a moving body as a series of points. Newton was arguing that it should be seen as a graph made by a continuously moving point. Since the velocity of a point moving in the direction x is the distance moved divided by the time, t, that is, x/t, interesting things can happen, he suggested, if we shrink both x and t. Thus a continuous and finite motion is equal to the quotient of an infinitesimal distance and an infinitesimal time. He gave the name *fluent* to the moving point, and he used the term *fluxion* for its velocity, that is, its derivative or rate of change with respect to time.

Publish or . . .

Had Newton been working today, he might have published something quickly in, say, the *Bulletin of the London Mathematical Society* and then perhaps more fully in a journal like Princeton's *Annals of Mathematics*. There he would likely start out by giving credit to the mathematicians on whose work he had built. Then he would explain his new work clearly, pointing out where and how he had moved ahead. In this way, he would clearly establish his priority, for the earlier publications would have been in peer-reviewed journals.

Unfortunately, there were at the time no such journals. This type of journal developed slowly and did not come into being until around the mid-1800s. Its objective appears to have been less to share new discoveries with the scientific community—there were already such journals in existence—than to provide a solid route to establishing one's priority in his or her discovery.

What did happen is that in 1669, Newton wrote up his early work as a tract, which he called *Analysis with Infinite Series* (often shortened to *De Analysi*), but it circulated only in manuscript form among a few colleagues, including Isaac Barrow, his teacher at Cambridge. It could, of course, have been published early on in book form—still a common form of establishing priority in Newton's day—but was not, for several reasons.

First, there was a severe recession in the book trade following the Great Fire of London in 1666, and technical works were at a particular disadvantage. Barrow, ironically, was in some measure to blame, for the publisher of his work had gone bankrupt, and so book publishers were particularly leery of publishing mathematical works.

Even so, things might have turned out differently but for yet another turn of fate's wheel. Newton was basically a loner. We have already seen that by age 23, and while still a student, he had already gone past the leading mathematicians of his day, and none but a few of his correspondents knew it. In 1669, thanks in part to his unpublished manuscripts, he was made Lucasian Professor of Mathematics at Cambridge University, which gave him the time and the freedom to continue his work.

He turned to his other interests, which included his first great discoveries in light and color, which had also been made in those remarkable years of the mid-1660s. Always hesitant to open his work to outside criticism, he nevertheless decided to try it, and he did so with a paper on this work in the *Philosophical Transactions of the Royal Society of London* in 1672. Although the paper received a generally good reception, Newton also found himself devoting precious time to answering sometimes inane challenges to his claims, always a danger when new ideas are presented. Among those objecting, unfortunately, were some eminent scientists, including the Dutch physicist Christiaan Huygens and the British scientist Robert Hooke. He found Hooke's criticisms especially troubling and distasteful.

As a result, although Newton continued to work on optics, he published no more papers on it and held off on publishing his major work on it, his *Opticks*, until after Hooke died—more than 30 years later! It may well be that this experience was a factor in his decision not to open his mathematics to the world. He seemed to believe that his discoveries belonged to him and not to the world, to science, or even to posterity. He may also have chosen to keep his discoveries close his chest to give him further time to refine them.

Whatever the reason or reasons, it was a decision that would cause him major problems, and mathematical historians considerable uncertainty, in the years to come.

By the 1680s, however, Newton had developed his work on mechanics, gravitation, and the movement of bodies to the point where he decided, after strong urging by his friend and colleague Edmund Halley, that he would put it into print. He began serious work on what would become his most famous work, the *Mathematical Principles of Natural Philosophy*, commonly known as the *Principia*, in

1684–1685. Published in 1687, it would go on to become possibly the most important and best-known publication in scientific history.

In it, he gave just a hint of the new calculus. He may have used the method to solve some of the problems he tackled in the book, then recast them and presented them in classic geometric fashion—perhaps to keep his calculus methods a secret for a while longer, but also because those were the standard methods of demonstration and proof.

Among these solutions was a conclusive demonstration that Cartesian vortices could not account for the planetary motions. Nevertheless, it took many decades before the authority of Descartes gave way to Newton's gravitational view of the universe.

Up to his *Principia* period, Newton's few contacts with Leibniz had been, on the whole, quite respectful and even friendly, but now he saw something in print that, if it didn't cause an immediate break, was surely a factor in what was to come. Before we get to it, however, we need to know something about its author.

Leibniz

Leibniz, born in 1646, was four years younger than Newton. Like Newton, he read and was influenced by Descartes' *Geometry*, as well as by other mathematical works. Yet even more, his interest in mathematics was stimulated by his earlier readings in philosophy. By the age of 6, he was already reading widely in the library of his father, a professor of moral philosophy at the University of Leipzig. By the age of 14, he was well-read in all areas of the classics.

Strangely, although he came from a middle-class family, among all the mathematicians/scientists/philosophers of his time he was the only one who had to scrabble for a living. That and his wide-ranging mind led him into a surprising variety of fields. By the age of 26, he had already designed a calculating machine that could add, subtract, multiply, divide, and even take roots; he had devised a program of legal reform for the Holy Roman Empire; and he had presented a plan to Louis XIV that involved a French attack on Egypt as a way of weakening the Ottoman Empire and deflecting French aggression away from Germany. Nothing came of it. At various times he was also interested in, and made contributions to, religion, philosophy, philology, logic, economics, and, of course, science and mathematics.

Herein lay a major difference between him and Newton. Newton was mainly interested in using his mathematics to solve scientific problems. Leibniz, like Descartes, hoped to make a major contribution to philosophy and thought that mathematics would pave the way. He wanted to create a kind of alphabet of human thought, in which symbols could be used to represent fundamental concepts, which could then be combined into more complex thoughts—a kind of calculus of reasoning.

Yet whatever Leibniz did in the field of mathematics, it was as a sideline to his multihued career, which makes his accomplishments all the more amazing. In 1673, he visited London on a diplomatic mission as part of his position as adviser to the archbishop of Mainz. There he met Henry Oldenberg, the secretary of the Royal Society, and made enough of an impression that he was elected to the Society. In other travels he had been in contact with the likes of men such as Huygens, Spinoza, Malpighi, and Vincenzo Viviani, a prominent pupil of Galileo's.

During his 1673 visit with Oldenberg, he may, says one historian of mathematics,[4] have seen a copy of Newton's *De Analysi*, though that seems unlikely. Even if he did, he might not have understood it. In 1676, he traveled to London, again as part of his diplomatic duties, and this time he visited with another colleague of Newton's, John Collins, who we know for sure showed him some of Newton's papers.

It was at this point that direct relations began between the two men. Leibniz was probably just beginning to think about the calculus, and the general feeling is that he was not only well behind Newton, but apparently did not even know of Newton's work in the area. And so, when Leibniz wrote to Newton, which he did twice in 1676, it was to ask questions about infinite series and their use in quadrature. Newton responded with two very respectful letters, which would play a strong role in the dispute that developed in later years.

While Newton's answers did skirt some issues of his calculus, he was careful to hide them in a carefully constructed anagram, or he simply alluded to such a method but never spelled it out. It was this disparity between their two stages that was to lead Newton to trouble, for when Leibniz did publish, some eight years later, Newton could not believe that Leibniz could have progressed so far so fast on his own.

Notation

Although Leibniz was deeply impressed with his own idea about a calculus of reasoning, it found little resonance among his contemporaries. More important by far, however, was that the mathematics that emerged from this work was to become the key to a far wider—and more directly useful—world of application. As with Newton's calculus, it became easier to deal with complex curves, areas, and volumes. Furthermore, it could deal handily with change—with velocities and accelerations, with rates of growth and decay—in ways that were just not possible before.

Finally, what both Newton and Leibniz had come up with was a method that did not merely provide solutions to a few specific problems, as earlier methods had, but an algorithm that had wide and spectacular generality. It could be applied to functions that were algebraic or transcendental (Leibniz's coinage), rational or irrational.

Over the development years, Newton used a variety of symbols, which caused some confusion later on. Early on, he tended to use the "little zero" to denote an arbitrary increment of time, and, say, op to denote the increment of a variable p. Later, he moved to the somewhat more familiar dot notation—for example, $\dot{x}$ for the first derivative of x (such as velocity) and $\ddot{x}$ for the second derivative (acceleration).

Leibniz was more careful and more thoughtful about his symbolism, and this would stand him in good stead when the scales of justice were balanced later on. For his differential calculus, he came up—after some trial and error—with the much more useful symbols dx and dy for the differentials (smallest possible differences) in x and y; and with the sign $\int$ for the integral function. For both men, finding tangents called for the use of the differential function, and calculating quadratures (areas bounded by curves) required the use of their integral calculus.

By the 1680s, then, Leibniz had established himself as an up-and-coming mathematician, and in 1684 his first description of his differential calculus was published in the journal *Acta Eruditorum*, under the title "A New Method for Maxima and Minima as Well as Tangents, Which Is Impeded Neither by Fractional nor by Irrational Quantities, and a Remarkable Type of Calculus for This." Here we see, for the first time, a clear statement of the basic formula for differentiation:

$$dx^n = nx^{n-1}.$$

In the same way that Newton did, he thought of integration not only as a summation of areas under a curve, but as the inverse of differentiation, and two years later he published his early work on the integral calculus.

It was with Leibniz's first publication, in 1684, that we see the first stirrings of trouble. Newton was not well known to the public but was well known and respected among his peers. He was beginning serious work on his *Principia*, and suddenly he was confronted with the first publication of the calculus. But it was by Leibniz—and there was no mention of Newton!

Was this unreasonable? Newton's reputation in mathematics was growing among his peers in England, but he still had absolutely nothing in print, and his name would have meant little to most Continental mathematicians.

In any case, he seemed utterly unmoved. In fact, he even acknowledged in the *Principia* that Leibniz had "fallen on a method of the same kind, and communicated to me his method, which scarcely differed from mine, except in notation and the idea of the generation of quantities."[5]

Other Players

Some of Newton's followers were less sanguine about this development. John Wallis, for example, felt that Newton's notions about fluxions were passing on the Continent by the name of Leibniz's differential calculus. By 1692, Wallis was putting together a collection of his work, and he strongly urged Newton to permit him to include something about Newton's calculus. The result was a mention of it in the preface to volume 1 of Wallis's *Works* (1695) and some excerpts in volume 2 (1693). (There is some uncertainty about the dating.)

Left alone, Newton and Leibniz might even have been able to remain on good terms. In March of 1693—nine years after Leibniz's first publication—for example, he wrote to Newton, trying to renew their correspondence, and though it took a while, Newton answered in October. His manner was still friendly. Certainly, there were no hints of anger or charges of plagiary in either man's letters.

Unfortunately, there were other players in the wings, even aside from Wallis, who would influence both men's behavior.

Neither Newton nor Leibniz had students to whom they passed on their work. After Leibniz had published his 1684 paper, however, the Swiss Bernoulli brothers, Johann and Jakob, had not only figured out the method but had already put it to use and passed it on to others. They also contacted Leibniz and began to act as his champions. Johann was especially active in this area, both directly and inadvertently. In the latter case, he set in motion a series of events that may well have been the precipitating cause of the heartbreaking feud that was to erupt.

In June of 1696, he issued a mathematical challenge to the "shrewdest mathematicians in the world": determine the curve linking any two points, not in the same vertical line, along which a body would most quickly descend from a higher to a lower point under its own gravity. He gave a private copy to Leibniz and also sent copies to Wallis and Newton. This was a clear challenge to Newton's method, and Newton did indeed solve it in a day. Newton sent his answer anonymously to the Royal Society. When Bernoulli finally saw it, however, he guessed at once that the author was Newton. He recognized, he said, the "the lion from his claw."

The answer is that it is a brachistochrone, which others had been able to figure out. The curve, however, is also in the form of a cycloid,* which could be understood only through use of the calculus. Leibniz

---

*The path taken by a point on the edge of a rolling disc. There is more on the brachistochrone problem in chapter 4.

then did a foolish thing. In 1699, he chose to write up a review of the solutions given earlier (May 1697) in the *Acta Eruditorum*, and he put forth the solutions as a successful demonstration of his own calculus. He also noted that there were a few others who had solved it, including Newton, but the implication was that all the others had used Leibniz's calculus. Thus Newton came out looking like a copier, or as a sort of pupil, of Leibniz.

Bernoulli, too, suggested that Newton, along with the others, was in some way indebted to the Leibniz/Bernoulli group. Here are the real beginnings of trouble. Aside from Newton's dot notation, said Bernoulli, there is little difference between the two calculus methods, and since Leibniz published first . . .

Neither Newton nor his English followers could have been happy about this, but there was one follower who was particularly annoyed. Nicolas Fatio Duillier was a Swiss mathematician who had moved to England and had become friendly with Newton. He had earlier worked with Huygens and had been a member of the Royal Society since 1687. Variously described as eminent mathematician, adventurer, prophet, mystic, and rogue, he took the omission of his name from the list of "eminent mathematicians" as a personal insult.

What should be done? The general reasoning of the Newton followers might have gone something like: at this late date, having lost out to Leibniz as far as first publication is concerned, it might be best to show that Leibniz's fame on the Continent is undeserved, that his formulation is inferior to Newton's and perhaps was even copied from him.

Duillier issued a lengthy analysis of the brachistochrone problem in a paper sent to the Royal Society; but, apparently quite annoyed with Johann Bernoulli and, by association, with Leibniz, he included therein some highly incendiary words about the origins of the calculus: "I am now fully convinced by the evidence itself on the subject that Newton is the first inventor of this calculus, and the earliest by many years; whether Leibniz, its second inventor, may have borrowed anything from him, I should rather leave to the judgment of those who had seen the letters of Newton, and his original manuscripts. Neither the more modest silence of Newton, nor the unremitting vanity of Leibniz to claim on every occasion the invention of the calculus for himself, will deceive anyone who will investigate, as I have investigated, those records."[6]

There it is. Without actually charging Leibniz with plagiary, Duillier has set down in print at least the possibility, if not the implication, of "borrowing" by Leibniz.

Things are heating up.

Was Duillier's attack made with Newton's connivance? There is no sure evidence either way. Was Newton sufficiently angry at this point that he would condone such an attack? Some say no, that he had not yet reached a boiling point. Others maintain that it is not likely that Duillier would have published such an attack without Newton's consent. We must leave it there.

More interesting is what came next. Leibniz was, of course, furious, yet he still felt that Newton might be innocent since Newton had given Leibniz credit for his work on the calculus in the first edition of his *Principia*. (Alert readers will wonder why I mention "first edition." Stay tuned.)

Leibniz had published a defense of his own activities and behavior in the *Acta*. To this, Duillier attempted to publish an answer, but the editors refused to accept it, on the grounds that the journal was no place for personal disputes. The dimensions of the feud are becoming clearer. Leibniz, it is worth noting, had helped to establish the journal and exerted some influence over its editorial activities, which might explain why his defense was published while Duillier's answer was not.

And there the feud lay for a few years.

Flashpoint

By the mid-1690s, Newton's interests were turning from science and mathematics (not to mention philosophy, religion, alchemy, and mysticism) to the political/administrative arena. In 1695, he contributed to discussions regarding reform of the country's currency, and a year later he was appointed warden of the Mint. This would require a move to London and a major change in his life in many ways. He continued his work on scientific subjects, but on a much-reduced scale, for he took his duties at the Mint very seriously. He was promoted to master of the Mint in 1699.

By now, he was hobnobbing with the wealthy and the powerful, aided by his attractive and vivacious niece Catherine Barton, who (most probably) lived with him in comfortable quarters in London.

He became more interested in the Royal Society and began to attend meetings regularly. He showed an improved design for a sextant, an instrument used for determining longitude at sea, but had another run-in with Hooke. In 1701, he read a paper on chemistry at one of the meetings, which was subsequently published. At the end of the same year he resigned his professorship at Cambridge, and, in recognition of his honored position, the university elected him one of its representatives in Parliament. He didn't do much there but continued his activities at the Royal Society.

In 1703, Hooke, whom Newton had tried so assiduously to sidestep, and who provided the reason why he had not published his work on optics for three decades, died. In the same year, Newton was elected president of the Society and was reelected each year until his death. Both events were to have profound consequences for both him and Leibniz.

True to his promise, Newton finally permitted his work on optics to be published. It was his second major publication, but this time, he apparently began to think of Leibniz as competition, because now he included in the book two papers on his calculus. One was "On Quadrature," which he had begun in 1691 and never finished. It finally appeared in 1704, but only as a supplement to his great book *Opticks*.

The "Quadrature" is interesting historically, in that it is to some extent a restatement and an expansion of the second 1676 letter to Leibniz and includes a translation of the fluxional anagram he had sent to Leibniz.

Further Conflict

To this point, we have looked at the situation mostly from Newton's point of view, but Leibniz, too, was enjoying a fast-growing reputation. In 1699, for example, the French Academy of Sciences created a list of eight Foreign Associates. Newton was seventh on the list; Leibniz was first. Newton's reputation, in other words, was still growing but far more slowly on the Continent than in Great Britain. Leibniz, on the other hand, may have been disturbed by the challenge posed by Newton; perhaps this lay at the heart of his next step.

A year later, Leibniz reviewed Newton's *Opticks*, anonymously, in the *Acta*. He called the book profound but found fault with the two mathematical supplements. Bad enough. Worse was a literary comparison that reminds us of the letters of Fermat in the previous chapter of this book. Leibniz stated that Newton had used his fluxions elegantly in his *Principia* and in other publications. No problem there. But then he added, just as Honoré Fabri in his *Synopsis of Geometry* had substituted progressive motions for the method of Cavalieri.

Cavalieri, a disciple of Galileo, was an excellent mathematician. Leibniz had in fact learned from Cavalieri's writings a technique for determining the area under a curve. Fabri, on the other hand, was a lesser light who had copied the thoughts of Cavalieri. The implication seems to be that Newton substituted his fluxions for the differences of Leibniz. Certainly, this is how Newton interpreted the comparison. He also immediately suspected Leibniz as being the author of these remarks and took this as an indirect but highly suggestive implication that Leibniz was the original inventor and that he, Newton, had somehow built on Leibniz's work or copied from it.

Could Leibniz really have meant to make such a direct attack? A. Rupert Hall, who has studied the matter carefully, argues, "I do not really think that Leibniz was expressing in this most sly and secretive fashion a hot and nourished resentment against Newton, as the latter came to suppose. Leibniz worked in great haste . . . [I]n this review he did not mean to bare his inward doubts and grievances; they slipped out, with fatal results." Hall concludes, "As wit wounds when laughter is intended, so did Leibniz's too-clever historical analogy."[7] Leibniz himself would later maintain that he had not meant to imply plagiarism.

In any case, there was an uneasy truce for a few years. Then, once again, a follower of one of the two men set the blaze roaring. This was John Keill, a Scottish student of James Gregory, who became Newton's next major champion. In what might otherwise have gone down as an unimportant paper in the *Philosophical Transactions* (October 1708), Keill included a statement about "fluxions which without any doubt Dr. Newton invented first, as can readily be proved by anyone who reads the letters about it published by Wallis; yet the same arithmetic afterwards, under a changed name and method of notation, was published by Dr. Leibniz in the *Acta Eruditorum*."[8]

Newton, apparently still not looking for direct confrontation, was initially annoyed by this publication. By some shrewd manipulation, however, in which Keill presented Leibniz's review along with his own paper to the Royal Society, Newton's anger was transferred from Keill back to Leibniz, and Keill's position as Newton's champion solidified.

For a variety of reasons, Leibniz didn't see Keill's article until 1710, and there was also some more back-and-forth activity. Perhaps with some prodding from Bernoulli, Leibniz then made a major tactical mistake. He appealed directly to the Royal Society for some sort of public exoneration. In a letter to the Royal Society (February 21, 1711), he protested that he had never heard the word *fluxions* before it appeared in Wallis's *Works*, nor was it to be found in Newton's letters to him of 1676; that in general the accusations were both "absurd and contemptible."

There are several reasons why this was a tactical mistake. To begin with, the feelings between the two men had been deteriorating for years. Now there was real anger on both sides. The problem for Leibniz was that he was walking directly into the lion's den. He was appealing to the very society of which Newton was the president. True, he had addressed his letter to Hans Sloane, the secretary of the Royal Society. What was the likelihood that Newton would stay out of the process? Furthermore, the British government was bitterly hostile to the threatened Hanoverian succession, with which Leibniz was connected. The very name Royal Society should have given Leibniz pause. Anything connected with the Hanoverians had a bitter taste for the Society's members, who tended to be Newton's supporters.

All of this was surely instrumental in Newton's decision to give Keill his full support in building a case against Leibniz. Excuse me. I meant to say, in setting up an official and unbiased commission of the Society, with the objective of investigating and answering the charges brought by Leibniz. It became, however, the occasion for Newton and his colleagues to amass a huge quantity of evidence, which was summarized and published as a report in February 1713, under the title *Commercium Epistolicum*. It was published anonymously, but those in the know could see Newton's hand in the process.

It hardly seems necessary to say that the report came down on Newton's side.

Later that year, Leibniz responded with the *Charta Volans*,[9] another supposedly anonymous publication. In it, he belittled the *Commercium* and directly accused Newton and his followers of stealing the differential calculus of Leibniz and, furthermore, of committing serious errors in their use of it.

Things, in other words, were getting worse instead of better. Now Keill published an article in the May–June 1713 issue of the *Journal Literaire de la Haye*, a French literary magazine, so the battle was leaking out onto public ground. In the same year, Johann Bernoulli published some mathematical criticisms of Newton's *Principia*. Bernoulli also referred to some of Newton's comments as "twice cooked cabbage."

Not to be outdone, Leibniz published his own *History and Origin of the Differential Calculus* (1714), spelling out his own answers to claims by the British mathematicians that he had taken his methods from Newton.

Then an "Account" of the *Commercium* was published, again anonymously, and again by Newton, in the Society's *Transactions*, in 1714.[10]

Among other claims, we find Newton's attempt to show that his calculus is superior to Leibniz's: "It has been represented that the use of the letter o is vulgar, and destroys the advantages of the differential method; on the contrary, the method of fluxions, as used by Mr. Newton, has all the advantages of the differential and some others. It is more elegant, because in his calculus there is but one infinitely small quantity represented by a symbol, the symbol o. . . . It [his calculus] is more natural and geometrical. . . . Mr. Newton's method is also of greater use and certainty. . . . When the work succeeds not in finite equations, Mr. Newton has recourse to converging series, and thereby his method becomes incomparably more universal than that of Mr. Leibniz, which is confined to finite equations."[11] And so on.

The feud was spreading in another way. Leibniz had, earlier (1710), criticized Newton's theory of gravitational attraction and its associated concept of action-at-a-distance, complaining that they smacked of the occult. Now (1716), lashing out again, he began to attack and even ridicule Newton's philosophical ideas. Newton felt, for example, that the universe could be thought of as a clock, one that God had wound up at the beginning of creation. Leibniz argued that if the clock ran on forever without God's help, what need is there for God?

Newton feared that some unexplained irregularities in the planets' motions might add up and finally throw the whole solar system out of kilter. God, he felt, would step in and set things right. Leibniz ridiculed Newton's idea of God as some sort of astronomical maintenance man. Leibniz argued that God would create the best possible world, for that is the nature of God. In addition, Leibniz and others had pointed to what they called the anti-Christian influence of Newton's *Principia*, which was worrisome.

The two men also had very different ideas on their concepts of space and time. Curiously, Leibniz's ideas were in some ways more modern than Newton's. It was essentially a clash between Newton's absolute concept of space and Leibniz's relational one, and we know who eventually lost out on that one. As for the solar system, Pierre Simon Laplace later proved that the solar system is stable. At the time, of course, Newton could only feel annoyed.

Again there were some back-and-forth exchanges, to little effect.

On November 14, 1716, Leibniz died. Was that the end of the dispute? No. Newton still felt the need to keep it going, as did a few of Leibniz's followers. In 1722, Newton arranged for a second edition of the *Commercium*. It was supposedly an exact reprint, but in Latin and with a few additions at the front. It was also, supposedly, edited by Keill, but Newton was really behind it. Because the original edition was hard to come by, this edition has become the basic reference for later scholars. It is a carefully reasoned document and presents Newton's case clearly and precisely—to the clear detriment of Leibniz.

There's just one small problem. A century later, when a scholar, Augustus De Morgan, compared the two editions, he saw clearly that Newton had changed, added, and omitted passages in the text—to his advantage, of course. The depth of Newton's anger begins to come clear. And then, some 12 years after Leibniz's death, when the third edition of the *Principia* was published, Newton removed all mention of Leibniz! As he argued, second inventors have no right. Thanks to his position at the Royal Society and his growing reputation as one of the foremost mathematical scientists of all time, he almost made that a true statement.

A Question Mark

Recall the modus of Duillier and the other followers of Newton. The best, and perhaps the only, way to counter Leibniz's growing reputation as the real inventor of the calculus was to demonstrate conclusively that Newton was first, that his calculus was superior, and to suggest that Leibniz may even have copied from him. We've seen the claim to the superiority of Newton's calculus in the *Commercium*. That was Newton's opinion, and he's entitled to his opinion—but there is more.

Note the date of Leibniz's first publication: 1684. Like Newton, he seemed in no rush to publish. Though he didn't wait almost 40 years, as Newton did, he did hold back for 9 years. Today's feverish rush to get into print seems not to have held sway at the time. Each man was, perhaps, hoping to further perfect the method before going into print.

Summarizing, then, Newton was surely first in its development: 1665–1666; Leibniz: 1673–1676. Leibniz, however, clearly published first: 1684–1686; Newton: 1704–1736.

Does this help us decide anything in terms of the priority dispute? Newton and his followers, of course, believed that he, being unquestionably the first to come up with the method, deserved all the credit, but the question is not really this simple.

First, Leibniz did publish first, and, as a result, his work was taken up and began to be applied before Newton's. Leibniz's notation was also superior and is the form we tend to use today. So posterity disagreed with Newton's claims of superiority.

As I noted earlier, Newton's first major publication in mathematics appears as a supplement to the *Opticks* book. Its overall title is

"Two Treatises of the Species and Magnitude of Curvilinear Figures"—work that he *may* have developed as early as the mid-1660s, but that did not see the light of day until 1704.

Similarly, Newton's next writeup of his technique, titled *Method of Fluxions and Infinite Series*, written in 1671, was not published until 1736 in English, and in Latin even later.

In other words, when Newton's publications began to appear, Leibniz had already become a threat. Newton and his followers could make a variety of claims, which may or may not have been true. They could say that Newton already had his calculus in hand by the middle and late 1660s; that he was already using the dot notation early on, whereas it may be that he actually did not begin using it until after he had seen some of Leibniz's work. Several historians maintain that Newton actually did not *begin* using his dot notation until the early 1690s.[12]

In the *Commercium*, Newton claims that he had written his Quadrature treatise as early as 1676, which it was later found was not so. He had actually written it in 1691.

Another claim in the *Commercium* is that Leibniz had seen a letter of Newton's dated December 10, 1672, concerning a problem on the tangent in which the method of fluxions was sufficiently described that "any intelligent person" would be able to come up with it. Later scholars agree that, first, any intelligent person would not be able to build a calculus on such flimsy hints; second, that Leibniz never did see the letter;[13] and third, that Newton knew, or should have known, this.

Wallis also contributed, perhaps inadvertently, to the barrage being launched against Leibniz. When he included material on Newton's calculus in his *Works* of the 1690s, he stated in his text that what he published was what Newton had sent to Leibniz in his letters, which was not the case at all. In fact, the collection he put together was not based on the original documents but on copies in which various passages were adapted to suit Newton's purposes.

In other words, much of what we "know" today about Newton's early work has come down to us via writings created *after* Leibniz had already become a threat. This doesn't automatically make them untrue, but they must be taken with a grain of salt.

Was There Plagiarism?

Rumors are given wings when there is both motive and plausibility. As the feud heated up, both sides were accusing the other side of plagiarism.

Certainly, there was motive on both sides. Both men felt that their reputations were on the line, and both had been prodded and poked by others until they were angry enough to do things that they might not have done—or permitted to be done—otherwise.

As for plausibility: Leibniz had seen some of Newton's papers. Furthermore, Newton had kept a few of his colleagues informed of his progress, some of whom were in contact with Leibniz. So Leibniz *could* have stolen from Newton. That in no way says he did.

The Newton side continued to complain about the fact that Collins showed Leibniz some of Newton's papers. Even here, though, it is not certain how much of Newton's calculus was contained in these papers. Current thinking is that there was very little, or at least that he used very little, and that what Leibniz saw and used had more to do with infinite series than with Newton's further work on the calculus itself.

Did Newton use the work of Leibniz? Here the plausibility scenario is even less certain. As we have seen, Newton and colleagues apparently had no compunctions about later changing facts to fit their own agenda. Yet after much thought and research by people who studied the matter in later years, the general consensus is that while both men can be accused of some impolite and even nasty conduct, neither was guilty of any form of plagiarism. That is, that each man came up with his calculus independently and without any direct input from the other.

Unexpected Outcome

It took a while for Newton's fame to spread, particularly on the Continent—but it did. At the same time, Leibniz's light seemed to go out, thanks to Newton's help, and toward the end of their respective lives, their situations were different indeed. Newton was idolized and had been knighted. When he died in 1727, he was given a state funeral, and he still lies buried in a prominent position in the nave of Westminster Abbey.

Leibniz's situation was very different. For him, nothing seemed to go right. In 1714, when the elector of Hannover, his employer, became George I of England, Leibniz even lost favor in his own court. This was almost surely because, amid all the diplomatic maneuvering, he was on the losing side of this important feud. In addition, he had tried to get the Roman Curia to release Galileo's *Dialogue* from the Index, also without success. He had hoped to help unify the Catholic and Protestant Churches, with obvious lack of success.

When Leibniz died in Hannover in 1716, unfulfilled in his many schemes, it seemed that he had hardly a friend in the court where he had labored for almost four decades. His funeral was attended by no one other than his former secretary. A friend noted in his memoirs that Leibniz "was buried more like a robber that what he really was, the ornament of his country."[14]

As a sort of final blow to the poor man's name, the French satirist Voltaire would do some serious lampooning of Leibniz with his *Candide* in 1759. It was to be Voltaire's most famous single work. Although a savage satire on 18th-century life and thought, it took particular aim at Leibniz. While the hero of the story is Candide, his mentor is Dr. Pangloss, a disciple of Leibniz. In spite of an extraordinary set of seriocomic misadventures, Pangloss maintains, as did Leibniz, that all is for the best in this best of all possible worlds. (Leibniz certainly spouted the best-possible-world part.) Voltaire was a devoted advocate of Newtonian ideas and did much to help spread them on the Continent. Recall, too, Leibniz's hope that his calculus might be able to unlock the secrets of human behavior. Voltaire ridiculed this idea as well.

Who Deserves the Credit?

Does Leibniz deserve any credit? Newton thought not, particularly in his later years.

There Newton was wrong, on at least two counts. First, he may not even have realized that the calculus he had come up with was a true advance. In other words, it wasn't until Leibniz and his followers showed the way that Newton understood that they had a general method that could be widely applied.

More important, however, is what happened in the years that followed the dispute. Although it was no longer going on between the two men, it had important repercussions. To put it simply, the British mathematicians stayed loyal to their man and would use only his calculus and his notation. On the Continent, however, from the very time of Leibniz's first publication, Leibniz's followers—and particularly the two elder Bernoulli brothers—took his new mathematics in hand and put it to work.

There were thus two important results of the feud. One was a breach between the two sets of mathematicians that lasted until the 19th century, which prevented the benefits that might have come from intercommunication between the two.

The second is even more significant: the rapid strides taken by mathematicians on the Continent—based largely on Leibniz's calculus—far outstripped those of the British during all of the 18th century! Here is where it may finally be said that Leibniz lost the battle but won the war.

4

Bernoulli versus Bernoulli

Sibling Rivalry of the Highest Order

The Bernoullis were an astonishing Swiss family, from which came eight noted mathematicians over a span of three generations.

The two main characters in our story are the brothers Jakob and Johann. Jakob, born in 1654, was the fifth of ten children. Their father, a successful spice merchant, wanted Jakob to go into the ministry. He even studied for it for a while. But Jakob's real interest was in mathematics, which he studied on his own. By 1676, at age 22, he was tutoring other students in the subject, and by 1687 he had become a professor at the University of Basel. At about the same time, and very shortly after Leibniz had published his first papers on his calculus (1684 and 1686), Jakob was already delving into it. In 1690, Leibniz would say of him, "The devices of this [Leibniz's] calculus are

yet known to few people, and I do not know anybody who has understood my meaning better than this famous man."[1]

Johann, the tenth child in the Bernoulli family, was born in 1667.
He was twelve and a half years younger than Jakob and, to his
father's distress, proved unsuited to a business career. In 1685, he
began the study of medicine and even got a degree in it, but, like
Jakob, his heart was in mathematics. He began to study it privately
with Jakob, probably in 1687. After perhaps two years, he was
already a match for his older brother. These two were the first mathematicians to recognize the calculus's importance, to put it to use, and
to spread the word about its significance.

By 1691, Johann was teaching the new mathematics to Guillaume
François L'Hospital—an experienced and gifted mathematician. Using
Johann's lesson plans, L'Hospital went on to write the first systematic text in the calculus (*Analyse des infiniment petits*, 1696). Johann also
taught mathematics to Leonhard Euler, who himself went on to
become a giant in 18th-century mathematics. In fact, nearly all of the
half dozen or so major mathematicians of the time had been pupils
of one or the other of these two Bernoulli brothers. By an interesting coincidence, Johann also had as a student J. C. Fatio de Duillier,
whose brother Nicolas, as we already saw in the last chapter, was to
play a major part in the Newton-Leibniz feud.

More important to our story, Johann taught mathematics to his
own two sons, Daniel and Nicholas, both of whom went on to
become quite respectable mathematicians in their own right. In fact,
the tradition continued, and a third son of Johann's also became a
professor of mathematics, and then *his* two sons became active in the
fields of science and mathematics. Today there are half a dozen mathematical equations, theorems, or functions named after a Bernoulli.

It's easy to imagine the Bernoullis as just one big, happy family
and the two brothers as especially pleased with their accomplishments and teaching careers.

Well, that's not quite the way it went. For although Jakob and
Johann were both successful and busy, and kept up an almost continuous exchange of ideas with Leibniz, with other mathematicians,
and with each other, they also challenged, argued with, and sniped
at each other at every opportunity. This was sibling rivalry on a

grand scale, for Jakob could never accept the fact that his much younger brother actually became his equal and, in some ways, even surpassed him. And Johann—well, we'll see shortly how Johann responded to his "kid brother" position.

Some Background

By the early 1690s, Jakob had done more to formalize the integral calculus than Leibniz himself had, for Leibniz had treated individual problems and had not set down general rules on the subject. Along the way, Johann became more proficient, and then, with their strange way of goading each other, both brothers used the new mathematics as a tool to solve problems that had bedeviled mathematicians for years, or even centuries.

For example, in 1659 Christiaan Huygens (1629–1695), a Dutch mathematician and physicist, had sought the curve along which an object descending under the influence of gravity would take the same amount of time to reach the bottom, from whatever point on the curve the descent began. He showed geometrically that the curve was a cycloid. Huygens then used this concept for his design of a pendulum clock that would keep accurate time. This design is sometimes referred to as an isochrone or a tautochrone. Galileo had earlier come up with the idea of using a pendulum for a clock, and Leibniz had done some preliminary work on the mathematics.

In May 1690, Jakob published his analysis of the equal-time problem, based on the calculus, in the *Acta Eruditorum*. He did it by setting up the differential equation for this curve of constant descent. The curve, he showed, is that of a cycloid. In essence, he proved Huygens's result analytically. The paper is important for another reason: the term *integral* appears for the first time as a calculus term.

Proud of his success with the isochrone problem, Jakob then proposed in the same paper an allied problem: determine the shape of a flexible but inelastic cord hung between two fixed points at the same height. The problem had been worked on at least as far back as the 15th century by Leonardo da Vinci; Galileo had considered the problem and had speculated that the curve was a parabolic arc.

The Kid Brother

Thirteen months after Jakob's paper, several solutions to the problem appeared in the June 1691 *Acta Eruditorum*. They were for a curve called a catenary, and they were by Leibniz, Huygens—and Johann! There was an entry from Jakob, which he called an "Additamentum ad Problema Funicularium." In it, he stated that after the solution was given by his brother, he continued the research further into some variations of the problem, such as cases where the rope is of different thickness or weight, for which he gave solutions.[2] As we'll see, Jakob's entry was not strictly an answer to the original problem and could be construed in different ways.

Johann made much of the fact that he was able to solve the catenary problem while, he maintained, his own brother, his teacher, could not. That was in 1691. The state of their later relations is shown by a letter Johann sent in 1718, some 27 years later, to his colleague and friend Pierre Rémond de Montmort. In it, he still writes in deprecating fashion of his brother, who had died 13 years earlier, and we still hear the strains of their never-ending competition. Msr. Montmort had apparently been under the impression that Jakob had been able to solve the catenary problem; Johann would have none of that. He wrote:

> The efforts of my brother were without success; for my part, I was more fortunate, for I found the skill (I say it without boasting, why should I conceal the truth?) to solve it in full and to reduce it to the rectification of the parabola. It is true that it cost me study that robbed me of rest for an entire night. . . . [B]ut the next morning, filled with joy, I ran to my brother, who was still struggling miserably with this Gordian knot without getting anywhere, always thinking like Galileo that the catenary was a parabola. Stop! Stop! I say to him, don't torture yourself any more to try to prove the identity of the catenary with the parabola, since it is entirely false. . . . [T]he two curves are so different that one is algebraic, the other is transcendental. . . . But then you [Montmort] astonish me by concluding that my brother found a method of solving this problem. . . . I ask you, do you

really think, if my brother had solved the problem in question, he would have been so obliging to me as not to appear among the solvers, just so as to cede me the glory of appearing alone on the stage in the quality of the first solver, along with Messrs. Huygens and Leibniz?[3]

Johann had in fact seen the important difference in the solutions.

For the parabola, the simplest form of the standard equation in Cartesian coordinates is

$$y^2 = 4ax$$

which is clearly an algebraic equation.

For the catenary, Johann showed that the equation is transcendental:

$$y = (a/2)(e^{x/a} + e^{-x/a}).$$

Shortly thereafter, Johann solved the differential equation for the velaria.* Not one to exult in private, Johann went around boasting of his achievements.

Already, though, we are beginning to face the difficulties involved in teasing out the realities of the two brothers' respective claims and counterclaims. Although Johann's claims to have solved the catenary and the velaria are backed up by several writers,[4] others dispute both claims. For example, W. W. Rouse Ball, a respected historian of mathematics, argues that credit for both of these developments belongs to Jakob. In fact, he claims that Jakob's solution to the problem of the velaria, as well as his proof that the construction given by Leibniz of the catenary had been correct, are among his greatest discoveries.[5] In addition, several sources say that it was Jakob's solution that later proved useful in the design of such structures as suspension bridges and high-tension towers.[6] As I noted earlier, Jakob's contribution in the 1691 article does contain a treatment of variants on the catenary problem, as well as further generalizations, which suggests that he did have some understanding of the original problem.

---

*Curve for a rectangular sail filled with wind.

A possible explanation for the confusion is given by Florian Cajori, who maintained that Jakob tended to publish answers without explanations, while Johann gave in addition their theory.[7]

Well, no matter. It's quite likely that by this time Johann was trying to break out of the kid brother role and was boasting of, and perhaps exaggerating, his accomplishments. Who could stop him from maintaining that he had the solution before Jakob, and a better one at that?

Their Young Years

Rüdiger Thiele, a professor of mathematics at the University of Leipzig, argues that the negative attitude of the two mathematicians' father had an unfortunate effect on both of their personalities. He feels that Johann, however, as the kid brother, suffered the most. As time went on, Johann compensated by developing an enormous sense of self-importance, and he tried in every way he could to achieve fame. Yet he always found himself under the shadow of Jakob. As a result, he attempted to exaggerate his own importance.

Thiele even argues that Johann's emotional problems actually made it difficult for him to properly evaluate his *own* mathematical achievements.[8] As for who deserves credit for the brothers' various discoveries, Thiele points out that in the early days of their relationship they worked closely together, so it is sometimes hard to distinguish their respective contributions.[9] This could explain the confusion about the solutions to the catenary and the velaria I spoke of earlier. Similarly, while the *Encyclopædia Britannica* states that Johann exceeded his brother in the number of contributions he made to mathematics,[10] Thiele feels that Johann perpetrated so many untruths (*Unwahrheiten*) that the fame he did accrue is unwarranted.[11]

But he reveled in it. In 1701, Johann wrote to his father, "That I never got a letter from my father indicates that he preferred my brothers and had no affection for me. Am I not worthy of as much consideration as my siblings? . . . I would be grateful if you could tell me how they have earned this trust and affection from you, which you are depriving me of. I have placed myself under God's guidance because my father will not allow me to lead the kind of life I would wish to lead. So don't come to Basel and take my fame and say that you had anything to do with it."[12]

Jakob felt just as strongly about their father and worked under his own motto, which might be translated as, "I went against my father's will, and yet I am up there among the stars."[13]

In any case, by the beginning of the 1690s Johann was already a bona fide mathematician and had developed to the point where his claims and his actual accomplishments made him a real threat to Jakob's own need for celebrity. The seeds were being sown for a serious battle. Thiele and others feel that as the feud escalated, Johann was usually the instigator, but the feeling is by no means universal. The mathematics historian J. E. Hofmann states that Jakob was "self-willed, obstinate, aggressive, vindictive, beset by feelings of inferiority, and yet firmly convinced of his own abilities. With these characteristics, he necessarily had to collide with his similarly disposed brother."[14]

It was during this period (early 1690s) that Johann had his dealings with L'Hospital. Johann was in Paris in 1691, where he was already earning respect as a practitioner of the new mathematics. L'Hospital, quickly seeing its true importance and value, hired Johann as his private tutor. L'Hospital was well off, and he paid Johann well. Even after Johann returned to Basel, he continued the lessons by correspondence. The written lessons provided the substance for the first textbook in differential calculus–L'Hospital's *Analyse des infiniment petits* (1696)–and is the basis for his respected name in the field.

Although Jakob was smarting from Johann's boasting, both brothers went on carving out their own mathematical careers, while at the same time interacting between themselves and with other major mathematicians of their time. They also continued working with the new mathematics, applying it to a variety of problems. In 1694, for example, Jakob came up with the analysis for a fascinating eight-shaped curve that is often seen in mathematics classes today. It is commonly referred to as the lemniscate of Bernoulli.

The Relationship Changes

In 1695, Johann was offered a professorship at Halle and at the same time the chair in mathematics at Groningen, in the Netherlands. He accepted the latter, though not without some resentment toward Jakob—he would have preferred Jakob's position at Basel but knew that was not possible as long as Jakob held it. Furthermore, Jakob had begun doing some damage control; in retaliation for Johann's boasting, he went about terming Johann his pupil, who could only repeat what he has learned from his teacher.

Yet Johann now ranked as high as Jakob, and in June of 1696 Johann posed, first in *Acta Eruditorum* and then via a leaflet, the problem of the brachistochrone: determine the curve linking any two points, not in the same vertical line, along which a body would most quickly descend from a higher to a lower point under its own gravity. Intuitively, one might expect the answer to be a straight line, that is, the shortest distance between the two points. Galileo had already realized, however, that that was not so, but his speculation that the path would be the arc of a circle was also not correct. As I noted in the previous chapter, the solutions offered by several mathematicians were published together in the May 1697 *Acta*. The honored group consisted of the two Bernoullis, Leibniz, Newton, and L'Hospital. It's worth noting that L'Hospital needed Johann's help.

The methods used by the two brothers to solve the problem are particularly interesting, for they illustrate well the difference in their characters and abilities. In essence, Johann performed a kind of trick. His ingenious mind saw a connection between the path of quickest descent, that is, the mechanical problem at hand, and Fermat's principle of least time and its application in an optical problem. From Snell and Descartes, he knew what happened when a ray of light passes from one optical medium to another. He figured he could combine the refractive sine law (chapter 2) with the equation for the velocity of a body under gravity:

$$v = \sqrt{(2gy)}.$$

Johann then divided the vertical plane of the problem into a series of very thin horizontal strips whose material densities varied slightly from one to the next. Although the particle would travel in a straight line through each strip, its path would bend slightly as it traveled from one to the next, as in a series of optical media with slightly varying refractive indexes. The particle's path of least time was therefore equivalent to the curved path of a light ray whose direction changed infinitesimally as it passed from one layer to the next.

Then, as Johann wrote, "But now we see immediately that the brachistochrone is the curve that a light ray would follow on its way through a medium whose density is inversely proportional to the velocity that a heavy body acquires during its fall. Indeed, whether the increase of the velocity depends on the constitution of a more or less resisting medium, or whether we forget about the medium and suppose that the acceleration is generated by another cause according to the same law as that of gravity, in both cases the curve is traversed in the shortest time. Who prohibits us from replacing one by the other?"[15] By letting the number of layers go to infinity, he was able to derive the curve for the brachistochrone.

Jakob, on the other hand, worked out a method that was more geometric and that seemed at first more cumbersome—he constructed a curve and used it as a basis for the analysis. As he put it, "The problem can therefore be reduced to the purely geometric one of determining the curve of which the line elements are directly proportional to the elements of the abscissa and indirectly proportional to the square roots of the ordinates."[16] The resulting curve was the one being sought. Both men had shown, each in his own way, that the correct form of the curve was a cycloid! The advantage of Jakob's method was that it was both more direct and more general. That is, it provided some general rules for solving several other problems of the same kind.

Calculus of Variations

The mathematics historian E. T. Bell argues, "James[17] [that is, Jakob] Bernoulli's signal merit was his recognition that the problem of selecting from an infinity of curves one having a given maximum or minimum property was of a novel genus, not amenable to the differential calculus and demanding the invention of new methods. This was the mathematical origin of the calculus of variations."[18]

The calculus of variations is a sort of generalization of the calculus. It seeks to find the path, the curve, or the surface for which a given function has a stationary value. In physical problems, this is usually a maximum or a minimum.

Though Bell credits Jakob for his pioneering input on this form of the calculus, there is a range of thought on this question—yet another example of how controversy seems to swirl around everything connected with the Bernoullis. Morris Kline, a well-known scholar, agrees with Bell, saying that while both solutions turned out to be early steps forward, Jakob's solution was even stronger in this regard.[19]

Yet Johann's solution, with its dependence on Fermat's principle of least action, surely points in this direction. Others feel even more strongly that the credit should go to Johann. David Eugene Smith, the editor of an important sourcebook in mathematics, writes, "The calculus of variations is generally regarded as originating with the papers of Jean [that is, Johann] Bernoulli on the problem of the brachistochrone." Smith's argument centers about the fact that Johann "attained a fairly complete if not precise idea of the simpler problems of the calculus of variations *in general*."[20]

The mathematics historian Stuart Hollingdale feels even more strongly: "It was Jean Bernoulli who started Euler on his researches into the calculus of variations."[21]

One of the problems here is that the question apparently hinges on which of Johann's solutions is being considered—and once again, the situation is cloaked in fog. J. J. O'Connor and E. F. Robertson, who did a series of articles on the Bernoullis for an online mathematics history forum, argue that Johann later built an elegant solution, published in 1718, that used a work of Brook Taylor's.[22] Smith maintains that this is not so, that "such a direct solution is mentioned in several of the letters which passed between Leibniz and Johann in 1696 as well as in the remarks which the former made on the subject of the brachistochrone problem in the *Acta Eruditorum* for May, 1697."

Smith admits that "this solution was not published until 1718 when both Jacques and Leibniz were dead." But he argues: "This fact is apparently regarded by those who believe Jean plagiarized from his brother Jacques as invalidating the former's claim of having secured a second solution. Jean for his part asserted that he delayed the publication of his second method in deference to counsel given by Leibniz in 1696."[23]

In rebuttal, Hollingdale argues, "However, until Euler took up the subject, no general methods were available."[24] In other words, the Bernoullis had used the method to solve specific problems, such as the brachistochrone. Euler, who began his work in this area around 1732, was more interested in a general theory. But the form in which the work is seen today was the work of yet another great mathematician, Joseph-Louis Lagrange.

What was Lagrange's take on the origins of the calculus of variations? Smith argues that he [Lagrange] "emphasizes the part of Jean no less than that of Jacques in pioneering work on a general method in the calculus of variations."[25]

And so we are pretty much back where we started. Well, let's go on.

Hollingdale adds an interesting point: "The development of the calculus of variations received a strong boost from physics, from the adoption by the eighteenth-century scientists of the 'principle of least action' as a guiding principle in nature."[26]

Ironically, the principle also had strong theological support. Euler stated, "For since the fabric of the Universe is most perfect and the work of a most wise Creator, nothing at all takes place in the Universe in which some rule of maximum or minimum does not appear."[27] Once again, Johann's use of Fermat's principle of least action comes to mind.

In any case, when the Bernoullis found that the cycloid was also the solution of the brachistochrone problem, they were amazed and delighted. Johann wrote in his article, "With justice we admire Huygens because he first discovered that a heavy particle falls down along a *common cycloid* in the same time no matter from what point on the *cycloid* it begins its motion. But you will be petrified with astonishment when I say that precisely this *cycloid*, the *tautochrone* of Huygens, is our required *brachistochrone*." Later, he picked up the notion again:

> Before I conclude, I cannot refrain from again expressing the amazement which I experienced over the unexpected identity of Huygens's *tautochrone* and our *brachistochrone*. . . . For, as nature is accustomed to proceed always in the simplest fashion, so here she accomplishes two different services through one and the same curve, while under every other hypothesis two curves would be necessary, the one for oscillations of equal duration, the other for quickest descent. If, for example, the velocity of a falling body varied not as the square root but as the cube root of the height [fallen through], then the *brachistochrone* would be algebraic, the *tautochrone* on the other hand [would be] transcendental.[28]

At the end of Jakob's paper on the brachistochrone, he laid out three other kinds of problems that can be solved by his method, the third of which was, "To find isoperimetric figures of different kinds." The origin of this problem may date back to pre-Greek times. In essence, it seeks to find which closed plane curve with a given perimeter will have the largest area. Within this group, Jakob crafted a complicated example and challenged Johann by name. He even offered Johann a prize of 50 ducats if he could solve it by the end of the year, or six months from then.

Now the fur really began to fly.

Battle Lines

Johann came up with a solution in 1697 and claimed the award. He failed, however, to perceive the isoperimetric problem's variational character and thereby offered an incomplete solution, one in which the resulting differential equation was one order too low. Jakob, delighted, criticized his brother unmercifully.

E. A. Fellman and J. O. Fleckenstein, writing in the *Dictionary of Scientific Biography*, argue, "This was the beginning of alienation and open discord between the brothers—and also the birth of the calculus of variations."[29] (Yet another variation on the origin of the calculus of variations.)

Johann's analysis of Jakob's isoperimetric problem was presented via the French mathematician Pierre Varignon to the Paris Academy of Sciences in February 1701. Jakob presented his solution in the *Acta Eruditorum* in May 1701. A later comparison of it with Johann's solution clearly shows Jakob's to be superior. Unfortunately, Jakob could not revel in this particular triumph. For reasons unknown, Johann's solution was put into a sealed envelope and not opened until April 1706, almost a year after Jakob's death.

Was this because Johann realized the truth even then? He never admitted it. Much later, though—after his brother's death, and after having been able to digest work by Brook Taylor (*Methodus Incrementorum*, 1715)—he produced an elegant solution of the isoperimetric problem. The concepts in this 1718 paper contain elements of the modern concept of the calculus of variations, which Euler and Lagrange would carry forward. The solution was, however, strangely reminiscent of Jakob's solution and style.[30]

Does Johann deserve the credit here? It would seem more appropriate to say that both men contributed—in their wondrously contentious fashion—possibly starting with the original solution of Jakob's.

Even in Death

Jakob's death in 1705 produced yet another instance of the strange relationship between the brothers. Among Jakob's many interests was the topic of probability, which he had pursued fairly intensely during the years 1684 to 1690. And while most of the mathematical contributions of the brothers were to be found in journals, especially in the *Acta Eruditorum*, Jakob spent the last two years of his life working on a manuscript for a text on probability. This was the *Ars Conjectandi*, or *The Art of Conjecture*.

It would contain a general theory of combination and permutation: the so-called weak law of large numbers—also known as Bernoulli's theorem and today used as a main tool in the theory of probability—and much else. It was his most important single piece of work. It was also the first substantial work on probability, and it still has application wherever statistical methods are used today, as in insurance, weather prediction, and population sampling.

The book is set up in four sections, the second of which, on permutations and combinations, he used for a proof of the binomial theorem for positive integral exponents. Contained in this section is a formula and a table for the sum of the rth powers of the first n integers. Using a table of his so-called numbers of Bernoulli, he calculated the sum of the 10th powers of the first 1,000 integers. Then, showing his sweet nature, Jakob wrote:

> With the help of this table it took me less than half of a quarter of an hour to find that the tenth powers of the first 1000 numbers being added together will yield the sum
>
> 91,409,924,241,424,243,424,241,924,242,500
>
> From this it will become clear how useless was the work of Ismael Bullialdus [which he] spent on the compilation of his voluminous *Arithmetica infinitorum* in which he did nothing more than compute with immense labor the sums of the first six powers, which is only a part of what we have accomplished in a single page.[31]

When Jakob died, the manuscript was nearly completed, but even in death, the animosity between the brothers played a role. It would seem logical for the work to have been published under Johann's supervision; but Jakob's widow was categorically against the idea, fearing that the vengeful brother might use the opportunity to damage or even sink the project. Nicholas, Johann's eldest son, had read the manuscript when he was studying with Jakob and, in true family spirit, had used it for his own thesis after Jakob died, and for other purposes as well. When it was finally published in 1713, it included a short preface by Nicholas. After admitting that he had been too young and inexperienced to do much with it, he says he advised the printers to give it to the public as the author had left it. It went on to become the centerpiece of Jakob's considerable reputation.

Jakob seems to have foreseen, or at least feared, an early death.

In the course of his studies, he worked with the curious equiangular spiral. This is a curve that can be seen in sea shells and the spider's web. It has some similarity with the circle, but there is one major difference. A circle crosses its radii at right angles; the equiangular spiral also crosses its radii at a constant angle, but not at 90 degrees. Jakob, who had some mystical leanings, was fascinated by the fact that the curve reproduces itself under various mathematical transformations. He asked that the curve be engraved on his tombstone, along with the inscription "*Eadem mutata resurgo*" (Though changed, I arise again the same). He died at the still young age of 51 in 1705, having held the chair in mathematics at Basel until his death. His chair at Basel was now available; it was offered to Johann, who was happy to accept it.

Johann Carries On

With Jakob out of the way, it was almost as if the argumentative Johann needed to find others to battle with. A healthy, vigorous man, he had another forty-three years in which to do so. L'Hospital's calculus text, for example, had been published in 1696. Initially, Johann seemed quite pleased with the way things had worked out. Upon receiving a copy from L'Hospital, he wrote back and expressed his thanks for being mentioned. He even promised to return the compliment when and if he, Johann, published such a work. L'Hospital actually suggested a follow-up, namely and reasonably, a text on integral calculus, since Leibniz didn't seem to be doing anything along these lines. Bernoulli, however, replied that he was, unhappily, preoccupied with domestic problems—which we'll get to in a moment.

That was Bernoulli's initial reaction, but the book had provided entry to a new and exciting world for Continental mathematicians and was received eagerly by them. As it became increasingly successful, Johann began to show an equivalent increase in jealousy and annoyance. In the years following Jakob's death, he complained bitterly about the situation. He attacked both the work and its author, virtually accusing L'Hospital of plagiarism. L'Hospital had acknowledged Johann's part in the work in the preface. He wrote, "And then

I am obliged to the gentlemen Bernoulli for their many bright ideas; particularly to the younger Mr. Bernoulli, who is now a professor in Groningen."[32] Johann now felt this was not sufficient credit, however, and he tried with all his might to tell the world who the real author was. But his cries seemed to fall on deaf ears.

For example, section 9 presents "Solution of some problems . . . ," involving what we would now refer to as indeterminate forms. Although the presentation is mainly geometrical, the result is what came to be called L'Hospital's rule—a mathematical method for evaluating indeterminate forms. This was particularly annoying to Johann, who felt that L'Hospital should have made it clear in the text that this was Bernoulli's work, not his. To be fair, though, L'Hospital never actually claimed this to be his invention. It was only by a quirk of fate, namely, the book's widespread use, that the result was so named—not by him but by others.

L'Hospital, it might be worth noting, was no longer around to defend himself. He had died in 1704. Nevertheless, and this would aggravate Johann even further, the *Analyse* remained the standard text for higher mathematics throughout Johann's long life—he died in 1748, at age 81—and even well beyond. A later comparison between Johann's lecture notes and L'Hospital's work showed them to be virtually identical, though a number of the mistakes in the notes do not appear in the book. So L'Hospital—or someone—did do some useful and knowledgeable editing.

According to one scholar, Gerard Sierksma, Johann's financial agreement with L'Hospital meant that Johann had sold his discoveries to L'Hospital and therefore could not publish his own work, at least for a while.[33] This could also explain Johann's unhappiness.

What were the domestic problems Johann referred to in his reply to L'Hospital? There were several that he might have had in mind. In 1697, he had lost a beloved daughter. Not long after, he suffered a serious illness.

More likely, it had to do with the years he was spending at Groningen, which ran from 1695 to 1705. Thanks to animosity between the city councilors and the provincial legislators because of religious differences, it was a trying time for everyone at the university. Johann had been brought up as a strict Calvinist and had remained a fervent member of the church, yet at least partly because of his work, he was accused of grievous heresies. Recall that Johann had also had medical training, and he had been expected to perform some medical duties as well. It seems that in the course of these duties, he made some comment about the continuous metabolism of the human body. He was attacked bitterly by both a student and a well-known theologian, Paulus Hulsius, and accused of denying the resurrection of the body.[34]

As he often did, he defended himself by going on the attack: "I would not have minded so much if he [the student] had not been one of the worst students, an utter ignoramus, not known, respected, or believed by any man of learning, and he is certainly not in a position to blacken an honest man's name and honour, let alone a professor known throughout the learned world, and distract the young from their fine studies."[35]

He was also attacked for espousing the use of experiments to learn about nature. He came through all of this, but there were some worrisome times.

Still More Controversy

Another man who can be counted among Johann's numerous adversaries was Brook Taylor (1685–1731). Taylor, in his *Methodus Incrementorum* of 1715, worked through many of the problems that Johann and others had dealt with, but the only credit Taylor gave was to Newton—not surprisingly, an Englishman. Bernoulli did not like being ignored and published an anonymous essay that accused Taylor of plagiarism. Taylor figured out who the author was and published an answer in defense—also anonymously. He also made fun of a mathematical error that Bernoulli had made years earlier. Johann's colleague, Pierre Rémond de Montmort, tried to mediate the dispute but got nowhere. The battle raged on until 1719, when Taylor published another insulting diatribe and then decided that enough was enough; Johann, characteristically, would have continued the battle.

Even years later, when Taylor died in 1731, Johann commented, "Taylor is dead. It is a kind of fate that my antagonists died before me, all younger than I. He is the sixth one of them to die in the last fifteen years. . . . All these men attacked and harrassed me . . . though I did them no wrong. It seems that heaven would avenge the wrong they have done me."[36] Note his reading of the turns of fate's wheel. The prior deaths of his adversaries proved, at least to him, that he was in the right!

As it turned out, neither Johann nor Taylor was aware that their battle was rather pointless. The general feeling today is that Taylor was not guilty of plagiarism, but rather of not keeping up with new developments on the Continent. Furthermore, the young Scotsman James Gregory had come up with the "Taylor series," which is basic in the method, some 40 years earlier.

And Yet Again

With the next controversy, we circle back to the family. Johann's second son was Daniel, who was born in Groningen in 1700. Johann, repeating his father's strange behavior, tried his best to prevent Daniel from pursuing mathematics. According to the mathematics historian James R. Newman, Johann went so far as to attempt to destroy the child's self-confidence through cruel treatment.[37] He tried, as his father had tried with him, to push Daniel into the business world, but something in the Bernoulli genes objected. Of course, in those days a young man was not likely to say, "Sorry, Pop, that's not for me. I want to study mathematics."

So, Daniel was first sent off as a commercial apprentice. When that didn't work, Johann sent him to study medicine, and he eventually earned his doctorate in 1721. His heart, though, as with so many other Bernoullis, was in mathematics. Historians differ over whether Johann taught Daniel any mathematics. If he did, it wasn't a great deal, and Daniel learned most of what he was taught from his older brother, Nicholas. As was the case with Johann, however, it was as a sideline. Nevertheless, he went on to become by far the ablest mathematician of the younger group of Bernoullis. By 1724, he had earned a solid reputation in the field with his *Exercitationes Mathematicae*, which covered several different mathematical subjects.

This led to a position in mathematics at the St. Petersburg Academy. He was, however, a scientific polymath, and he worked not only in mathematics but also in such fields as medicine, biology, physiology, physics, mechanics, astronomy, and oceanology. The following year he won a prize awarded by the Paris Academy, the first of 10 he would earn in these various fields.

Nicholas had also gotten an appointment at St. Petersburg. But within eight months of their appointments, Nicholas died, leaving Daniel feeling both lonely and not very happy with the harsh climate. Johann once again entered the scene, and we see an excellent example of his complicated personality. Argumentative, irascible, and jealous, he nevertheless arranged for one of his best pupils, no less a one than Leonhard Euler, to move to St. Petersburg to work with Daniel in 1727. The following few years were among Daniel's most productive. One of his main topics of interest was vibrating systems; by 1728, Daniel and Euler were doing important work on the mechanics of flexible and elastic bodies.

Ironically, all of Johann's three sons gained some reputation as mathematicians and scientists, but Daniel was to become the most famous of all. When Johann began feeling the hot breath of competition from this youngster, he reacted badly to it.

The battle between the optics and the dynamics of Newton and the once-dominant Cartesian description of the world was still being fought, including at the Paris Academy of Sciences. Johann, arguing against Newton's ideas, had won the Paris Academy's prestigious prize competition twice, in 1727 and 1730. He won again in 1734, but this time he had to share the prize with Daniel, who was arguing in support of Newton!

Daniel had been wanting to come back to Basel for years. He had finally obtained the chair of anatomy and botany at Basel in 1734. He had come home. Unfortunately, Johann was increasingly unhappy with having to share credit with his own son. The occasion of the shared prize led to a break between them; Johann even barred Daniel from the family home.

Then things went from bad to worse. Both men continued to work in mathematics and its application to physical problems. Daniel had been working on a text titled *Hydrodynamica*, which covered the properties important in fluid flow, including pressure, density, and velocity. It included the key relationship, now called Bernoulli's principle, which states that pressure in a fluid decreases as its velocity increases. He had the manuscript ready by about 1734, but for various reasons it was not published until 1738. This was to be Daniel's most important work, and the work that would make him famous.

His father, not to be outdone, published a competing book, which he titled *Hydraulica*, at about the same time. Again, history is somewhat ambiguous. At the very least, Johann tried to predate his book to 1732, to make it seem like Daniel took his material from Johann.[38] At worst, he actually stole material from Daniel's book and tried to pass it off as his own, and then attempted to make it look as if Daniel had stolen from him![39] Later, in a 1743 letter to Euler, Daniel referred to "my complete *Hydrodynamic*, for which I do not have to thank my father," and complained, "I was robbed of and lost the fruit of my ten years of labor."[40]

This hardly sounds like the behavior of a proud and loving father, yet this was the same man who had thoughtfully sent Euler up to St. Petersburg to help Daniel through a difficult time. In any case, fate would serve up another unpleasant dish to Johann, for he lived another 10 years, long enough to see Daniel's work become a classic text in the field.

Why end on a sour note, though? Daniel was a first-class scientist, as well as a competent mathematician, and he has been described as one of the founding fathers of mathematical physics. If Johann couldn't find pleasure in his son's success, that was his problem. He had two other sons, who also became competent mathematicians, and several grandsons, with the same outcome. The only real blight I can see is that the eldest of his three sons, Nicholas, died at Petrograd, where he was a professor, at the young age of 31. Yet Galton describes Nicholas as "a great mathematical genius" and "one of the principal ornaments of the then young academy."[41] Who can say what Johann's feelings really were?

In any case, Johann, like his brother Jakob, and like Daniel, had plenty to feel good about. We've seen that Johann and Jakob were the very first to see the importance of the new calculus, and that both of them performed some very useful mathematics during their time on earth. After the death of Jakob in 1705, Leibniz in 1716, and Newton in 1727, Johann reigned as perhaps the foremost mathematician of his time. It was largely due to his efforts—both through his teaching and through his many demonstrations of the calculus's wonderful powers—that the differential notation of Leibniz, rather than Newton's fluxional notation, was generally adopted on the Continent. He was surely one of the great teachers of all time. He was also, it turns out, one of the great communicators. His scientific correspondence adds up to some 2,500 letters, and he exchanged letters with no fewer than 110 scholars.

5

Sylvester versus Huxley

Mathematics: Ivory Tower or Real World?

Thomas Henry Huxley, a highly respected 19th-century British scientist, made important contributions in zoology, geology, and anthropology. One of his biographers, Adrian Desmond, writes of the era, "By 1870 science was Professor Huxley."[1]

Yet this was an era when schools and universities stressed theology and the classics; science was a rich man's pastime. In 1870, Oxford gave out 145 classics fellowships and just 4 in science.[2] Huxley's lectures and writings were instrumental in convincing academics and policy makers that this was a mistake. He stressed the importance of scientific training as a way of bringing the mind "directly into contact with fact, and practicing the intellect in the completest form of induction; that is to say, in drawing conclusions from particular facts made known by immediate observation of Nature."[3]

An accomplished speaker and writer, Huxley also delved deeply into religion—though hardly in any conventional manner. The eminent botanist Sir Joseph Dalton Hooker wrote in a letter to Charles Darwin, "When I read Huxley, I feel quite infantile in intellect."[4]

Yet somehow Huxley's brilliance and wide-ranging intellect seemed to stop short at the world of mathematics. As he saw it, mathematics is virtually all deductive and therefore does not provide the same kind of valuable training of the mind that science does. He felt it was more of a game, that it was, along with classics and theology, not really on the same level as scientific training and ability. Hence, he could write, "The mathematician starts with a few propositions, the proof of which is so obvious that they are called self-evident, and the rest of his work consists of subtle deductions from them."[5]

As for its connection to science, mathematics "knows nothing of observation, nothing of experiment, nothing of induction, nothing of causation."[6] It is, in short, useless for scientific purposes. He stated these ideas at a scientific meeting, where they might have gone unremarked, but then wrote them in articles in two popular journals, *Macmillan's Magazine* and *The Fortnightly Review*.[7]

People other than Huxley had had such, or similar, thoughts before and had even stated them. Students have long complained, for example, does it really matter that the three angles of a triangle are equal to two right angles? Or that every even number greater than 2 is, or may be, the sum of two primes? Henry John Stephen Smith, a fine mathematician and a contemporary of Huxley, proposed a toast at a banquet: "Pure mathematics; may it never be of use to any one."[8]

Was Smith serious? Only if you *define* pure mathematics as those aspects of the field that are not useful—at the moment. Smith surely knew of examples of pure mathematics that were later put to work, as, for example, when Apollonius's work on conic sections was used 19 centuries later by Kepler in his investigations of the orbits of the planets.

Nor does it matter. Such comments never had much effect. But when a scientist of Huxley's stature and reputation voiced ideas like this, and to a broad public, more than one mathematician was more than a little put out, and the mathematical world knew someone had to give a solid response. The man they chose to give this response was James Joseph Sylvester, an eminent British mathematician. The occasion was the Presidential Address to the Mathematical and Physical Section of the 1869 meeting of the British Association, also known as the British Association for the Advancement of Science.

Even though Sylvester was by then one of the major algebraists of his day, it required some courage—some might even say foolhardiness—to take on a person of Huxley's eminence and fighting spirit. By then, Huxley's defense of the newly created, and widely hated, evolution doctrine had earned him the nickname Darwin's Bulldog. Desmond states, "A pugilistic fame put Huxley in the papers almost weekly."[9]

Before we consider Sylvester, his challenge, and the results, let's see if a closer look at Huxley's upbringing and training will provide any insight into his feelings about mathematics.

Huxley before Sylvester's Challenge

Huxley, born in 1825, lived above a butcher's shop in a sleepy village called Ealing. His father was an assistant schoolmaster at a school there—and taught mathematics! The school was said to be one of the finest private schools in England.

Huxley attended this school, but his father left the school in 1835 for unknown reasons. This ended Huxley's primary schooling, with the result that he had only two years of formal instruction in his early years. Yet he had this to say of his time there: "The society I fell into at school was the worst I have ever known. . . . the people who were set over us cared about as much for our intellectual and moral welfare as if they were baby-farmers. We were left to the operation of the struggle for existence among ourselves, and bullying was the least of the ill practices current among us."[10] Perhaps this had something to do with Huxley's later habit of lashing out, often mercilessly, at opponents and ideas that displeased him.

It's not clear where his feelings about mathematics came from. He showed early signs of brightness in other subjects, however, and by his teens was thinking hard about philosophy and was already developing a classification system for knowledge. By age 17, he had divided all knowledge into two major groups, objective and subjective. He wasn't sure where to put morality, though he was rather inclined to place it under the objective heading. Into that category went physics, physiology, and history. Under subjective, he grouped metaphysics, theology, logic—and mathematics!

If he was not a math wizard, though, early studies in medicine stood him in good stead in various other fields that he turned to later on. By age 29, he had written 23 papers on invertebrate zoology and had 15 more in progress. He also made important contributions in taxonomy, which earned him election to the Royal Society in 1851 and the Royal Medal in 1852.

More generally, he felt that without some knowledge of science, often called natural history in his day, one could not be called educated. In fact, he felt, "To a person uninstructed in natural history, his country or sea-side stroll is a walk through a gallery filled with wonderful works of art, nine-tenths of which have their faces turned to the wall."[11]

He wrote that in 1854. At the same time, his feelings about mathematics were already hardening. In the same essay, he stated:

> I do not question for a moment, that while the Mathematician is busy with deductions *from* general propositions, the Biologist is more especially occupied with observation, comparison, and those processes which lead *to* general propositions. All I wish to insist upon is, that this difference depends not on any fundamental distinction in the sciences themselves [that is, between biology and mathematics], but on the accidents of their subject-matter, of their relative complexity, and consequent relative perfection.
>
> The Mathematician deals with two properties of objects only, number and extension, and all the inductions he wants have been formed and finished ages ago. He is occupied now with nothing but deduction and verification.
>
> The Biologist deals with a vast number of properties of objects, and his inductions will not be completed, I fear, for ages to come; but when they are, his science will be as deductive and as exact as the Mathematics themselves.[12]

In 1856, Huxley would still state, "The mathematician discovers in the universe a 'Divine Geometry.'"[13]

In the various worlds of science, however, Huxley was still learning and growing.

The biographer James G. Paradis writes:

> The drift of Victorian science . . . was toward an emphasis on quantification and empirical determinism. In . . . 1854, Huxley had furnished a short summary on scientific method. . . . By 1870, when the essay was published as part of his *Lay Sermons*, his point of view had so changed that he took special care to note that he no longer adhered to theories of vitalism. *Lay Sermons* contained his "On the Physical Basis of Life," which rejected all theories of spontaneity and identified life forces with chemical forces. While Huxley had drawn from the Romantics for his early concepts of nature, [by 1866] he publicly abandoned his earlier Romantic tendencies, rejected spontaneity, and declared scepticism to be "the highest of duties."[14]

We begin to see more clearly how Huxley viewed science and mathematics. He loved science and had matured in understanding and dealing with it. In his mind, science was part of life. One could not be truly educated without a good grounding in its various parts but especially in natural history or natural science. As to mathematics, he did not hate it; he admired it, but preferably from afar. He saw it as some sort of game, perhaps a wonderful game, but disconnected from science. This even though, as Paradis points out, science was moving toward quantification.

In 1868, he clarified his position a bit further: "But the man of science, who, forgetting the limits of philosophical inquiry, slides from these formulae and symbols into what is commonly understood by materialism, seems to me to place himself on a level with the mathematician, who should mistake the x's and y's with which he works his problems, for real entities—and with this further disadvantage, as compared with the mathematician, that the blunders of the latter are of no practical consequence, while the errors of systematic materialism may paralyse the energies and destroy the beauty of a life."[15]

Note Huxley's expression "the blunders of [mathematics] are of no practical consequence." Then consider that in our own day, an $18.5 million space probe—the Mariner 1 mission of July 22, 1962— was lost because a hyphen was missing from a computer equation.

Here's more from the same 1868 paper. We see here his thinking about mathematics being purely deductive. Note, too, how he compares the learning of a language with that of mathematics. In the third paragraph, he describes the state of education in his day.

> If the great benefits of scientific training are sought, it is essential that such training should be real: that is to say, that the mind of the scholar should be brought into direct relation with fact, that he should not merely be told a thing, but made to see by the use of his own intellect and ability that the thing is so and not other-wise. The great peculiarity of scientific training, that in virtue of which it cannot be replaced by any other discipline whatsoever, is this bringing of the mind directly into contact with fact, and practising the intellect in the completest form of induction; that is to say, in drawing conclusions from particular facts made known by immediate observation of Nature.
>
> The other studies which enter into ordinary education do not discipline the mind in this way. Mathematical training is almost purely deductive. The mathematician starts with a few simple propositions, the proof of which is so obvious that they are called self-evident, and the rest of his work consists of subtle deductions from them. The teaching of languages, at any rate as ordinarily practised, is of the same general nature—authority and tradition furnish the data, and the mental operations of the scholar are deductive. . . ."
>
> At present, education is almost entirely devoted to the cultivation of the power of expression, and of the sense of literary beauty. The matter of having anything to say, beyond a hash of other people's opinions, or of possessing any criterion of beauty, so that we may distinguish between the Godlike and the devilish, is left aside as of no moment. I think I do not err in saying that if science were made a foundation of education, instead of being, at most, stuck on as cornice to the edifice, this state of things could not exist.[16]

Darwin's Bulldog

Over the course of Huxley's long career, there were many people who felt the sharp tip of his active pen. He feared not reputation or position. Many of his battles had to do with Darwin's theory of evolution. Darwin, ironically, had neither the heart for nor the interest in defending against the inevitable attacks, and he feared the worst. Early on, Huxley tried to calm him: "And as to the curs which will bark and yelp, you must recollect that some of your friends . . . may stand you in good stead. I am sharpening up my claws and beak in readiness."[17]

One of Huxley's most famous bouts took place in 1860, the year after Darwin had sprung his theory on the world. Huxley took on the bishop of Oxford, Samuel Wilberforce, known to a select few as Soapy Sam. Coached by Richard Owen, one of the major scientists of his day, Wilberforce was expected to chop Darwin's evolutionary theory—and by extension, Huxley—to pieces. It was Huxley's job to stop him. This he did in a fine theatrical performance, and it helped earn him the title of Darwin's Bulldog.[18] I mention it only to emphasize Huxley's willingness to take on all kinds of challenges, and challengers of all stripes and sizes.

Another important battle fought by Huxley, of quite a different kind, took place almost a decade later. By the mid-1800s, uniformitarianism had become the dominant geological theory. It suggested that the past could be explained in terms of the geological actions and forces we see acting today. The main importance of uniformitarian theory was that there was no need for such catastrophes as the Flood or for any supernatural influences; it thereby seemed to offer an effective refutation of the Christian idea of a very young earth, shaped by catastrophes.

Because uniformitarian theory called for these earth-shaping forces to be acting over unlimited time, it fitted in nicely with Darwin's ideas; in fact, Darwin was influenced by the idea. Unhappily, both uniformitarianism and Darwinian evolution were theories, and difficult to prove. In addition, they were faced with a challenge that was virtually impossible to answer. It came from a group of mathematical physicists, the most important being William Thomson. A major force in 19th-century physics and mathematics, and later to become Lord Kelvin, Thomson had done some excellent work on thermodynamics, the scientific application of heat and work. Based on this work, and on some very careful mathematical calculations, Thomson came to the conclusion by the mid-1800s that the earth could be no more than 400 million years old.

If Thomson was right—and no one could find anything wrong with his calculations—then several major theories, including uniformitarianism and Darwin's evolution, were unworkable. So solid was Thomson's reputation that the evolutionists even tried to shorten the time needed by evolution to do its work, to no avail.

A few of Thomson's opponents, including both uniformitarian geologists and biologists, accepted the accuracy of his calculations but began to question his assumptions, and so, some nine years after Huxley had debated Bishop Wilberforce, Huxley once again found himself chosen to do battle in a major arena. This time, however, the debate was to be held in a more scientific arena, the Geological Society of London. More important, his opponent was to be a far more capable adversary. This time it was William Thomson—who, by the way, had attended the earlier Huxley-Wilberforce debate.

Although the two men found themselves debating in some very deep waters, including the origins of life on earth, the verbal debate settled nothing. Yet the back-and-forth challenges carried over into the following years and drew in other entries as well.

In a review of Huxley's lecture in the *Pall Mall Review* of May 3, 1869, John Tyndall called it "one of the most able addresses ever delivered by a President of the Geological Society."[19]

On the other hand, Albert Ashforth, another of Huxley's biographers, writing with the benefit of hindsight, later called Huxley's performance at the debate "his most unconvincing performance as a defender of Darwin."[20] Perhaps, but Huxley made an interesting point. "I desire to point out," he declared, "that this seems to be one of the many cases in which the admitted accuracy of mathematical processes is allowed to throw a wholly inadmissible appearance of authority over the results obtained by them [scientists]. Mathematics may be compared to a mill of exquisite workmanship, which grinds you stuff of any degree of fineness; but, nevertheless, what you get out depends upon what you put in; and as the grandest mill in the world will not extract wheat-flour from peascods, so pages of formulae will not get a definite result out of loose data."[21]

Perhaps he was suspicious for the wrong reasons. No matter. He was surely correct in his suspicion of Kelvin's figure. The figure was far too low, but the reason was not to be uncovered for another quarter of a century. A correct understanding had to await the discovery of radioactivity—the heat from which Thomson had not taken into account in his calculations, and this had thrown him off considerably.[22] Huxley's intuitive statement had been right on the mark.

We see that Huxley would attack any idea, popular or not, if he felt it was wrong. An issue that he found particularly troubling was a growing interest in the Positivist philosophy of Auguste Comte. Though Comte had died in 1857, his followers were pushing the idea that they were sympathetic to the sciences, yet Huxley felt that implicit in their beliefs was an authoritarian spirit that was completely antithetical to his own ideas on the intellectual freedom needed in the sciences. He was also unhappy with attempts by Comte's followers to see, or create, connections between religion and science.

Huxley would have none of that, and he decided the only way to deal with this would be to show that Comte didn't have a clue what science was all about. So, when others were giving high praise to Comte's Positivist philosophy, Huxley directed some sharp criticism at it. In February 1869, he published a critical article in the *Fortnightly Review*. Among a variety of stinging comments, we read, "What struck me was his [Comte's] want of apprehension of the great features of science; his strange mistakes as to the merits of his scientific contemporaries; and his ludicrously erroneous notions about the part which some of the scientific doctrines current in his time were destined to play in the future. . . . [I]t has been a periodical source of irritation to me to find M. Comte put forward as a representative of scientific thought."[23]

What could have caused Huxley to be so acerbic? Surely, one reason was Comte's ideas on mathematics. In the same *Fortnightly Review* article, he quotes Comte as follows:

It is therefore by the study of mathematics, and only by it, that we can get a fair and profound idea of what is a science. It is uniquely there, where one must seek to know with precision the general method which the human mind uses constantly in all of its positive inquiries, because nowhere else are questions resolved in such a complete manner and deductions taken as far with rigorous severity. It is also there that our understanding has given the greatest proofs of its force, since the ideas that are considered there are of the greatest degree of abstraction possible in the positive order. All scientific education which does not begin by such a study [of mathematics] is necessarily lacking a solid base.[24]

"That is to say," Huxley concludes triumphantly, "the only study which can confer 'a just and comprehensive idea of what is meant by science,' and, at the same time, furnish an exact conception of the general method of scientific investigation, is that which knows nothing of observation, nothing of experiment, nothing of induction, nothing of causation! And education, the whole secret of which consists in proceeding from the easy to the difficult, the concrete to the abstract, ought to be turned the other way, and pass from the abstract to the concrete."[25]

Huxley was aware of the adversarial role he was playing in these and many other controversies. He was often attacked, for example, for his rather freethinking ideas on orthodox religion, as well as for his defense of Darwinism. These never bothered him, and, as we've seen, when he thought it necessary, he could answer with powerful argument and stinging rebuke.

So it must have been something of a surprise, and a major shock, when a famous scientist stood up at an important scientific meeting and said that Thomas Henry Huxley may know a great deal about science, but when it comes to mathematics, which is so intimately tied to science, he does not know what he is talking about.

And in the Other Corner

I remember a joke that made the rounds in the early days of the civil rights era. A Southern redneck breaks his arm and goes to the only black doctor in town to get it set. The redneck's friends are aghast that he would go to a black doctor. "What'd ya do that for?" they ask. The redneck responds, "If he can become a doctor in this town, he must be good."

Something similar can be said for Sylvester, for he was a practicing Jew at a time and a place where this was not an easy thing to be, particularly if there was something special about him.

Born James Joseph in London on September 3, 1814, he was the sixth of nine children. An older brother had gone to the United States, where he had taken the name Sylvester, a renaming followed by the rest of the family. What could have possessed observant Jews to take on a name associated with early Christian popes who certainly had no love for Jews? Perhaps they thought it would somehow ease the path of the younger Sylvesters in pursuing their lives and vocations. If that was the objective, it certainly did not work for James Joseph Sylvester.

Sylvester didn't talk much about his early years, though we know he attended private schools between the ages of 6 and 14. His mathematical talent must have shown itself early, for the last half of his 14th year was spent at the University of London (later University College), where he studied under the brilliant mathematician/teacher/writer Augustus De Morgan. He was expelled after an altercation with a fellow student whom he threatened with a table knife. We do not know how much provocation there was on the other side.

At age 15 he entered the Royal Institution at Liverpool. At the end of his first year there, he won a prize in mathematics and was so far ahead of his classmates that he was placed in a class by himself. Yet his two years at Liverpool were not happy ones. Open about his religion, he was persecuted constantly by his so-called Christian classmates. He finally called it quits and fled to Dublin with only a few shillings in his pocket.

From 1831 to 1837—with a two-year break due to illness—he attended St. John's College, Cambridge, and graduated as second wrangler, a high honor. Yet he was refused a degree because, as a Jew, he would not subscribe to the so-called Thirty-Nine Articles prescribed by the Church of England as a requirement for obtaining the degree. He had to go to Trinity College in Dublin, where he was granted a B.A. in 1841.

Sylvester had other such experiences, but in all cases he fought back. The mathematics historian Eric Temple Bell maintains—with a touch, perhaps, of poetic exaggeration—that Sylvester "was no meek, long-suffering martyr. He was full of strength and courage, both physical and moral, and he knew how to put up a devil of a fight to get justice for himself—and frequently did. He was in fact a born fighter with the untamed courage of a lion."[26]

In 1838, he found his first job. It was at University College, London, where he was to teach natural philosophy, not mathematics. A major difference between Huxley and himself began to show up early, for Sylvester proved unsuited to the teaching of science in general and physics in particular, even though his old professor, Augustus De Morgan, was one of his colleagues and tried to help him.

Karen H. Parshall, who edited a collection of his letters, states, "He disliked mounting experiments for his classes, and he proved virtually incapable of drawing diagrams on the board, despite lessons from the college's drawing master aimed at overcoming the problem."[27] Huxley, on the other hand, had always been a fine illustrator, which had served him in good stead at his work. Such differences do not, of course, guarantee antipathy, but we can perhaps begin to see why and how they viewed the world so differently.

In the meantime, other aspects of Sylvester's remarkable career were falling into place. He began to make regular contributions to the *Philosophical Magazine* and, in April of 1839, was elected a fellow of the Royal Society at the unusually young age of 25 for his work in "Physical and Mathematical subjects," that is, in applied mathematics.

Sylvester stayed only two years at University College and in 1841 tried his luck in the United States, at the University of Virginia. This position lasted just three months, ending amid conflicting reports of a run-in with a student and/or the administration. In one version of the entanglement, the administration refused to back him up in his punishment of a student who had insulted him.[28] In another, he had to leave because of his outspoken antipathy to Negro slavery.[29]

There are other versions as well. They all, however, seem to involve some sort of conflict.

After unsuccessful attempts to find other employment in the United States, he returned to Europe in the middle of 1843 and decided to leave academic life, at least for a while. He continued, however, to give some private lessons in mathematics and even had as a student Florence Nightingale, who was later to become famous as both a founder of modern nursing and a pioneer in sanitation and hygiene.

He did some work as an actuary and then turned to law for a while. He was called to the Bar in November 1850.

Although his fellowship at the Royal Society had been awarded for his work in applied mathematics, his concentration in later years was almost exclusively in pure mathematics. This interest was kindled mainly by his association with Arthur Cayley, whom he later met (around 1851) while both he and Cayley were working in the field of law. Yet both men went on to become two of the world's great mathematicians.

Indeed, even when working at law, both men found time for mathematics, and each inspired the other. From then on, Sylvester was reenergized and produced some of his best work.

In 1854, he was an unsuccessful candidate for the chair of mathematics at the Royal Military Academy, at Woolwich. Yet he achieved a kind of moral victory in the summer of 1855 when he obtained—with the aid of an influential supporter—an examinership at the academy and thereby forced, as he put it, "the practical admission (the first I think in this country) that a Jew as such shall not be debarred from public situations for which he is competent."[30]

The moral victory turned into more than that when the new incumbent conveniently died soon afterward, and Sylvester was named in his place. Sylvester wrote to his patron and supporter, Lord Brougham, thanking him profusely and promising to dedicate himself to both study and service.

Unhappily, Sylvester once again found himself at odds with one group or another. Parshall writes that he "viewed himself as a researcher and teacher (in that order); the authorities saw him as a teacher alone and had little sympathy for his desire for time to pursue his own researches. These conflicting expectations, coupled with Sylvester's strong and long-standing sense of faculty autonomy in classroom matters, provoked tension repeatedly throughout his fifteen-year tenure at Woolwich."[31] The result was that he was productive while at Woolwich but had to fight constantly for the time in which to prepare his papers and to take care of other matters. In 1855, he had become the editor of the *Quarterly Journal of Pure and Applied Mathematics*. In 1863, he became the mathematics correspondent to the French Academy of Sciences.

From 1866 to 1868, he was president of the newly formed London Mathematical Society. It was, all in all, a busy and productive, if difficult, time for Sylvester.

Over the long years of his career, he produced many papers on mathematics, including such areas as the theory of numbers, differential equations, and higher algebra.

He was both inventive and productive. His papers, published after his death, fill four large volumes. Though he worked in many areas, much of his work was in resolving basic problems in algebra; this included work on the roots of quintic equations. He worked on the calculus. Although his inventiveness was somewhat marred by his failure to include rigorous proofs for his ideas, he was adept at creating the language and the notation in these areas as they were needed. He coined much of the terminology in the theory of invariants, as well as the often used concept of the matrix, the rectangular array of numbers from which determinants can be formed. It was this sort of inventive genius that prompted the Royal Society to award him the Royal Medal in 1861 and, later, the Copley Medal in 1880.

Bell describes him: "Sylvester, short and stocky, with a magnificent head set firmly on broad shoulders, gave the impression of tremendous strength and vitality, and indeed he had both."[32] Exaggeration or not, Sylvester felt, as he himself wrote in later years, that he had been fighting the world all his days.[33] James R. Newman, the creator of the masterful four-volume *The World of Mathematics*, describes him as brilliant, quick-tempered, and restless.[34]

So Sylvester was a brilliant mathematician, as well as a fighter, and was known to be an excellent speaker. All of this gave the Council of the British Association good reason for choosing him as its champion. Yet Sylvester's willingness to enter the ring may also have played a part, for he may in fact have harbored some personal animosity toward Huxley.

A new club had been formed in November 1864. This was the X Club, a highly selective group of Royal Society members that virtually ran British science. Huxley was included; Sylvester was not. The small group met for dinner before the monthly meetings of the Royal Society and kept going until 1892 when it literally died of old age. Sylvester was a close friend of one of the members—namely, the mathematician and physicist Thomas Archer Hirst—and was on speaking terms with most of the other members. For whatever reason, though, he was never invited into the club, in spite of the fact that there were 9 members and it was initially intended to consist of 10. Furthermore, it did include a couple of mathematicians. Of all Huxley's various connections, this club was the one he valued the most, and considering Huxley's important position in British science, Sylvester may well have blamed Huxley for his not being invited in. He later referred to the group as a "cabal."

In any case, this was the man deemed best able to respond to Huxley's charges against mathematics.

Sylvester's Response

The occasion was the 1869 Presidential Address to Section A—the Mathematical and Physical Section—of the British Association. Sylvester had been elected president of this section, with, as he put it, the "tranquilizing assurance that it would rest with myself to deliver or withold an address as I might see fit."[35]

At first, Sylvester felt that he had no address to deliver. On reflection, however, he decided that "failing an address, the members would feel very much like the guests at a wedding-breakfast where no one was willing or able to propose the health of the bride and bridegroom."[36] This seems now like a curious explanation, sounding more like the introduction to a toast than a roast, and indeed Sylvester began his address with some very respectful remarks. For example, he referred to Huxley as "one whom I no less respect for his honesty and public spirit than I admire for his genius and eloquence." Then, however, he added, "But from whose opinions on a subject which he has not studied I feel constrained to differ."[37]

Sylvester has begun his attack. It builds slowly: "I have no doubt that had my distinguished friend, the probable President-elect of the next Meeting of the Association, applied his uncommon powers of reasoning, induction, comparison, observation, and invention to the study of mathematical science, he would have become as great a mathematician as he is now a biologist. . . . and the eminence of his position and the weight justly attaching to his name render it only the more imperative that any assertions proceeding from such a quarter, which may appear to me erroneous, or so expressed as to be conducive to error, should not remain unchallenged or to be passed over in silence."[38]

After a few more introductory remarks, Sylvester quoted Huxley's inflammatory statement: "'Mathematics is that study which knows nothing of observation, nothing of experiment, nothing of induction, nothing of causation.'"[39]

Sylvester's answer:

I think no statement could have been made more opposite to the undoubted facts of the case, [I think] that mathematical analysis is constantly invoking the aid of new principles, new ideas, and new methods, [is] not capable of being defined by any form of words, but spring[s] direct from the inherent powers and activity of the human mind, and from continually renewed introspection of that inner world of thought of which the phenomena are as varied and require as close attention to discern as those of the outer physical world . . . that it is unceasingly calling forth the faculties of observation and comparison, that one of its principal weapons is induction, that it has frequent recourse to experimental trial and verification, and that it affords a boundless scope for the exercise of the highest efforts of imagination and invention."[40]

Sylvester went on to give some examples to back up his claims, such as, "Lagrange, than whom no greater authority could be quoted, has expressed emphatically his belief in the importance to the mathematician of the faculty of observation; Gauss has called mathematics a science of the eye, and in conformity with this view always paid the most punctilious attention to preserve his text free from typographical errors."[41]

Then, moving to a theme that Huxley could hardly have imagined, Sylvester continued, "The ever to be lamented Riemann has written a thesis to show that the basis of our conception of space is purely empirical, and our knowledge of its laws the result of observation, that other kinds of space might be conceived to exist subject to laws different from those which govern the actual space in which we are immersed, and that there is no evidence of these laws extending to the ultimate infinitesimal elements of which space is composed."[42]

Later, he stated, "Most, if not all, of the great ideas of modern mathematics have had their origin in observation." Among the several examples he gave is "Sturm's theorem about the roots of equations, which, as he informed me with his own lips, stared him in the face in the midst of some mechanical investigations connected with the motion of compound pendulums."[43]

After several examples, he added:

I might go on, were it necessary, piling instance upon instance to prove the paramount importance of the faculty of observation to the process of mathematical discovery. Were it not unbecoming to dilate on one's personal experience, I could tell a story of almost romantic interest about my own latest researches in a field where Geometry, Algebra, and the Theory of Numbers melt in a surprising manner into one another, like sunset tints or the colours of the dying dolphin, "the last still loveliest" (a sketch of which has just appeared in the *Proceedings of the London Mathematical Society*), which would very strikingly illustrate how much observation, divination, induction, experimental trial, and verification, causation, too (if that means, as I suppose it must, mounting from phenomena to their reasons or causes of being), have to do with the work of the mathematician.[44]

Then, however, Sylvester did a curious kind of backpedaling and said the same sort of things about mathematics education in England that Huxley had been claiming for British education in general right along:

> I, of course, am not so absurd as to maintain that the habit of observation of external nature will be best or in any degree cultivated by the study of mathematics, at all events as that study is at present conducted[,] and no one can desire more earnestly than myself to see natural and experimental science introduced into our schools as a primary and indispensable branch of education: I think that study and mathematical culture should go hand in hand together, and that they would greatly influence each other for their mutual good. I should rejoice to see mathematics taught with that life and animation which the presence and example of her young and buoyant sister could not fail to impart . . . [and] the mind of the student quickened and elevated and his faith awakened by early initiation into the ruling ideas of polarity, continuity, infinity, and familiarization with the doctrine of the imaginary and inconceivable.[45]

Again, in a mode reminiscent of Huxley's complaints about British education in general, "It is this living interest in the subject which is so wanting in our traditional and mediaeval modes of teaching. In France, Germany, and Italy, everywhere I have been on the Continent, mind acts direct on mind in a manner unknown to the frozen formality of our academic institutions' schools of thought, and centres of real intellectual cooperation exist; the relation of master and pupil is acknowledged as a spiritual and a lifelong tie, connecting successive generations of great thinkers with each other in an unbroken chain."[46]

Then, however, he returned to his basic point: "When followed out in this spirit, there is no study in the world which brings into more harmonious action all the faculties of the mind than the one of which I stand here as the humble representative, there is none other which prepares so many agreeable surprises for its followers . . . , or

. . . seems to raise them, by successive steps of initiation, to higher and higher states of conscious intellectual being."[47]

Sylvester had to know that he was locking horns with an experienced, determined, and unusually disputatious competitor. Though trying not to insult directly, he had to know that he had nevertheless directly challenged Darwin's Bulldog, one who was known to have, in the words of Adrian Desmond, "a stiletto of a pen."[48]

Huxley after Sylvester

Yet somehow, and for whatever reason, Huxley never offered a direct reply to Sylvester's British Association lecture. Considering his apparently strong feelings on the matter, and his usual willingness to take on all challengers, this seems strange indeed.

There are several possibilities here. One is that he never saw or heard anything about Sylvester's charges. Considering both men's connection with the British Association, this is very unlikely.

Another is that Huxley was simply overworked. One of his biographers, Cyril Bibby, writes that for some time during the 1870s, Huxley "seems to have remained uncharacteristically non-combatant," explaining that "he was desperately overworked and chronically weary and dyspeptic."[49] This may just as well have been the case in 1869.

In any case, Bibby also points out that "on all the major issues of philosophy, Huxley's final positions did not differ greatly from those taken as a young man."[50]

Did this include Huxley's feelings about mathematics? One scholar, Alexander MacFarlane, argues in his *Lectures on Ten British Mathematicians of the Nineteenth Century* that Huxley, "convinced or not . . . had sufficient sagacity to see that he had ventured far beyond his depth."[51]

I wonder. With regard to Sylvester's charges, it may be that Huxley found little to argue with, and that he actually did change his thinking as a result. If there were such changes, however, they were subtle. I have seen little on this in his biographies or in his own later writings, but he does make occasional reference in these writings to mathematics.

For example, in an 1876 lecture giving his proposals for a good university education, he suggested that "Mathematics will soar into its highest regions."[52]

It sounds like he has begun to see the error of his thinking. Yet when he wrote in greater detail, a different picture emerges. In 1872, he felt that "In our ideal University, a man should be able to obtain instruction in all forms of knowledge. Now, by 'forms of knowledge' I mean the great classes of things knowable." He talked of three classes: the first had to do with our mental facilities, for example, logic, psychology, and so forth; the second concerned man's welfare and conduct, and included moral and religious aspects; but it is the third class that we are most interested in. He wrote, "A third class embraces knowledge of the phænomena of the Universe, as that which lies about the individual man; and of the rules which those phænomena are observed to follow in the order of their occurrence, which we term the laws of Nature.

"This is what ought to be called Natural Science, or Physiology, though those terms are hopelessly diverted from such a meaning; and it includes all exact knowledge of natural fact, whether Mathematical, Physical, Biological, or Social."[53] In other words, Huxley is still classifying mathematics under the rubric of exact knowledge that must be learned along with physical, biological, or social knowledge.

Similarly, in 1882 he could write, "But a great mathematician, and even many persons who are not great mathematicians, will tell you that they derive immense pleasure from geometrical reasonings. Everybody knows mathematicians speak of solutions and problems as 'elegant,' and they tell you that a certain mass of mystic symbols is 'beautiful, quite lovely.' Well, you do not see it. They do see it, because the intellectual process, the process of comprehending the reasons symbolised by these figures and these signs, confers upon them a sort of pleasure, such as an artist has in visual symmetry."[54]

Note his term *mystic symbols* and his comparison with the pleasure an artist derives from viewing something felicitous or beautiful. Of course, there is beauty to be seen in mathematics, but my guess is that Huxley was still blind to what it is that mathematics involves and offers that is beyond beauty.

My reading of these words is that, to put it bluntly, the mathematician is, in general, not really doing anything useful, as the scientist is. (As I mentioned earlier, this was not, and is not, a position unique to Huxley.)

Sylvester after Huxley

In 1870, the year after the British Association meeting, Sylvester was forced to retire from Woolwich at the ripe old age of 56. Much of his most important work was still in the future, but his superiors in the War Office assumed that a man of his age should retire. In the process, the school's administration tried to cheat him out of some of his retirement income, but he fought for it and won his case.

Though honors still came his way, he turned more to his beloved poetry, read the classics, and played chess. In other words, he thought that his productive days in mathematics were pretty much over. In 1876, however, he was called to Johns Hopkins University as its first professor of mathematics. Sylvester came to mathematical life once again and went on to spend seven glorious and productive years at the renowned American institution.

Ironically, the institution had opened not long before, and its keynote speaker had been another Englishman—Thomas Henry Huxley!

So both men were deeply involved in education, though, of course, from different points of view. Huxley, so far as I am concerned, was never fully convinced by Sylvester's arguments about mathematics.

Though Sylvester remained a British subject, he played an extremely important role in stimulating the mathematical research community in the United States, as well as in Great Britain. In his teaching, he was just as likely to have his students working on a new problem as trying to solve an old one. His teaching methods, his love for the subject, his enthusiastic and broad-based activity all helped to create a fine department of mathematics and a strong start for research mathematics in the United States—not only via the university but also by his founding and editing the *American Journal of Mathematics*, to which he contributed 30 papers. In addition, he also fought for the right of women to study in the field when it was not the usual thing to do.

Sylvester wanted to bring mathematics into peoples' lives, to make it a living thing. Huxley wanted to do the same for science. Science has become so much a part of our everyday world, and of our educational process, that we tend to forget that only a short century and half ago, there was strong opposition to the addition of science at both lower and higher education levels. The same holds for Darwin's theory of evolution. Today's much more balanced curriculum owes its existence in large part to Huxley's early urgings.

His view of mathematics remained cramped, however—rather like a set of rules that, once learned, said it all. Sylvester's view couldn't be more different. As he saw it,

> Mathematics is not a book confined within a cover and bound between brazen [brass] clips, whose contents it needs only patience to ransack; it is not a mine, whose treasures may take long to reduce into possession, but which fill only a limited number of veins and lodes; it is not a soil, whose fertility can be exhausted by the yield of successive harvests; it is not a continent or an ocean, whose area can be mapped out and its contour defined; it is limitless as that space which it finds too narrow for its aspirations; its possibilities are as infinite as the worlds which are forever crowding in and multiplying upon the astronomer's gaze; it is as incapable of being restricted within assigned boundaries or being reduced to definitions of permanent validity."[55]

Clearly, Huxley and Sylvester had very different ideas on mathematics. But with regard to British education, and the American educational system as well, we were lucky to have had both men as champions.

6

Kronecker versus Cantor

Mathematical Humbug

A variety of famous definitions, axioms, and notions lie at the heart of Euclid's geometry, a self-contained branch of mathematics that has inspired generations of mathematicians with its precision and order. One of Euclid's well-known "common notions" states that the whole is greater than the part. It stood without question for more than 2,000 years.

Then, in the early 1870s, an unknown mathematician began to claim that when it comes to numbers and number theory, the whole may *not* be greater than the part. It seemed an outrageous claim, and one that could have been safely ignored had it not been made by a brilliant, tenacious young German by the name of Georg Cantor. And that was only a small part of Cantor's math-shaking claims. Another had to do with that mysterious concept known as infinity.

Mathematicians and philosophers had long nibbled around the edges of this arcane concept. Some, like Galileo, argued that infinity is "by its very nature incomprehensible to us."[1] Karl Friederich Gauss wrote to another mathematician, "I protest against the use of an infinite quantity as an actual entity; this is never allowed in mathematics. The infinite is only a manner of speaking."[2] Cantor not only argued for a real, concrete infinity but insisted that there are different sizes and found a way to deal with the concept mathematically.

How important was his work? He created set theory, which became the basis for topology, fractals, and much else in our modern times. Cantor's work with sets led to advances that helped provide a rigorous grounding for the calculus. Combining the two concepts, set theory and infinity, he came up with infinite sets, an idea that was to create excitement for others, new research fields for many, and both pain and pleasure for himself.

Cantor's new theory of numbers would influence and challenge a generation of mathematicians in his own day. It led to a critical investigation that shook the very foundations of mathematics. In addition, its implications and paradoxes have continued to challenge the following generations of mathematicians right up to the present. The eminent German mathematician David Hilbert would later describe Cantor's work as "the most astonishing product of mathematical thought, one of the most beautiful realizations of human activity in the domain of the purely intelligible."[3]

It's not unusual for someone introducing new work to face opposition, but Cantor's treatment seems to have been especially severe and unpleasant. Joseph W. Dauben, possibly the foremost Cantor scholar in the United States, compares Cantor's treatment by his contemporaries with that of Giordano Bruno by the Inquisition. He writes, "While none of the major participants in the modern attempt to move from a closed mathematical universe into a surprisingly and complexly infinite one was ever burned at the stake, Georg Cantor, in a less dramatic way, faced inquisition and repudiation at the hands of many of his contemporaries."[4]

Foremost among these opponents was Leopold Kronecker, a highly placed and influential German mathematician. Kronecker had been one of Cantor's teachers; he had in fact been an early supporter of Cantor's work and had even provided constructive criticism for one of Cantor's early papers. As Cantor's work veered toward the unorthodox, however, Kronecker turned more and more against both Cantor and his work. At the height of their conflict, says Dauben, "Kronecker considered Cantor a scientific charlatan, a renegade, a 'corruptor of youth.'"[5]

Kronecker

Born in 1823 into a wealthy German Jewish family, Leopold Kronecker was given a good early education and had the luxury of being able to carry on his mathematical studies for as long as he wanted. At his hometown school in Liegnitz (now Legnica, Poland), he studied mathematics with Ernst Eduard Kummer, who went on to do brilliant work in higher arithmetic and geometry, and who would remain a close friend to Kronecker. In 1841, Kronecker enrolled at the University of Berlin, by that time the mathematics capital of the world, and studied with some of the best people in the field—P. G. Lejeune Dirichlet, Carl Gustav Jacobi, and Ferdinand Gotthold Eisenstein. He received his doctorate in 1845 but was diverted from a mathematical career for a full decade by pressing family business. In this time he married, began a family, and continued to do some mathematics but strictly as a hobby.

In 1855, he and his family moved to Berlin, and he began a professional career in mathematics, but then things moved quickly for him.

It was a time of important change in the German mathematical scene. The University of Göttingen persuaded Dirichlet to come to it from Berlin, and Kummer was brought in to fill the vacancy. At Kummer's suggestion, Karl Theodor Weierstrass—a successful secondary school teacher who had just published a well-received paper on power series representation of a function—was also brought on board. Although Kronecker was not yet a member of the Berlin faculty, he was producing papers at a good rate, and his growing reputation brought him a membership in the Royal Academy of Sciences in Berlin; this in turn carried with it the right to give lectures at the university.

In 1866, Kronecker was offered a good position at the University of Göttingen, but life was too good in Berlin, and he declined. Still, it was not until 1883, when Kummer retired, that Kronecker became a professor at the University of Berlin. Nevertheless, from the 1860s on, these three—Kummer, Weierstrass, and Kronecker—would be the ruling triumvirate in German mathematics for more than a quarter of a century. Kronecker was particularly active in the Berlin Academy, where he was instrumental in recruiting many of the most important mathematicians, both foreign and German. Among them were Sylvester and Richard Dedekind, of whom we'll have more to say later.

Kronecker was also a member of several other societies, and his advice was often requested when it came time to fill mathematical posts both in Germany and abroad. In 1880, he became the editor of August Leopold Crelle's *Journal for Pure and Applied Mathematics*. Often referred to simply as *Crelle's Journal*, it was probably the most respected mathematical journal of its day.

Kronecker's main achievements lay in his efforts to unify arithmetic, algebra, and analysis and in his work with elliptical functions. He introduced a number of refinements in algebra and in the theory of numbers, as well as many new concepts and theorems, such as his theorem on the convergence of infinite series.

Ironically, Kronecker was a bit of a maverick himself. He believed, for example, that all of arithmetic could be based on whole numbers. Thus he regarded fractional numbers as possessing only a kind of derivative character and as being useful only for notational purposes. He classified all mathematical disciplines except geometry and mechanics, but including algebra and analysis, as arithmetical. Because he believed that all of arithmetic could be based on whole numbers, he felt that not only fractions but irrationals and complex numbers as well were false or illusory ideas, and they had arisen through the application of some sort of false mathematical logic.

Thus, when Ferdinand Lindemann wrote a paper containing a proof of the existence of transcendental numbers, Kronecker would comment, "Of what use is your beautiful research on the number π? Why cogitate over such problems, when really there are no irrational numbers whatever?"[6] He believed that eventually a way would be found to recast these "unnatural" forms into a more elementary form involving only the natural numbers. A beautiful one-liner of his states, "God made the integers; all the rest is the work of man."

Is it any wonder that Kronecker and Cantor struck sparks?

But Kronecker's approach to mathematics also put him into disagreement with some of his contemporaries. Though Kronecker was careful not to get into a shouting match with people he saw as adversaries, or to commit nasty remarks to print, he was not above *behaving* in a nasty or hurtful way or saying damaging things behind their backs. Among others he went after was his one-time good friend Weierstrass. The two spent their last years in complete disagreement over their mathematical views. Kronecker was also annoyed and frustrated by Weierstrass's great success as a teacher.

We get an idea of how Kronecker worked from an 1885 letter Weierstrass sent to his colleague Sonya Kowalevsky. He wrote,

> But the worst of it is that Kronecker uses his authority to proclaim that *all* those who up to now have laboured to establish the theory of functions are sinners before the Lord. When a whimsical eccentric like [Elwin B.] Christoffel says that in twenty or thirty years the present theory of functions will be buried . . . we reply with a shrug. But [then] Kronecker delivers himself of the following verdict which I repeat *word for word*: "If time and strength are granted me, I myself will show . . . a more rigorous way. If I cannot do it myself those who come after me will . . . and they will recognize the incorrectness of *all* those conclusions with which *so-called* analysis works at present." Such a verdict from a man whose eminent talent and distinguished performance . . . I admire as sincerely . . . as [do] all his colleagues, is not only humiliating for those whom he adjures to acknowledge as in error . . . but it is a direct appeal to the younger generation to desert their present leaders and rally around him. . . . Truly it is sad, and it fills me with a bitter grief, to see a man whose glory is without a flaw

let himself be driven by the well justified feeling of his own worth to utterances whose injurious effect upon others he seems not to perceive.[7]

physical -1,

Kronecker was a small man; Weierstrass was a ~~big~~ large man. The mathematics historian Amir D. Aczel describes what he calls the "comic nature of these clashes, with the small man constantly attacking the large one like a small dog going after a St. Bernard."[8] Weierstrass even considered decamping for Switzerland to escape the constant bickering with Kronecker, but he feared that Kronecker would have considerable input in the choice of his successor—and that all his, Weierstrass's, work would be undone by anyone acceptable to Kronecker. He stayed on. By 1888, he let it be known to several of his friends that his friendship with Kronecker was finished. Kronecker, however, apparently never understood how much his behavior had affected Weierstrass and on several later occasions still referred to him as a friend.

Kronecker also had an interesting run-in with his longtime friend and colleague Hermann Amandus Schwarz. Schwarz was Kummer's son-in-law and had been Weierstrass's student. Weierstrass, remember, was a big man; Kronecker was not only very short, but was very self-conscious about it. In 1885 Schwarz sent him a greeting that included the statement: "He who does not honor the Smaller, is not worthy of the Greater." Schwarz apparently thought he was cleverly, and humorously, honoring Kronecker. Kronecker did not see the joke, but again, there was no sharp written or even verbal response. He simply had no further dealings with Schwarz.

So Cantor was not the only person with whom Kronecker had a falling out. Yet when we look at all the factors—his rigid ideas on numbers and infinity; his reputation and powerful positions in both the academic and the publishing world; and his ability and inclination to throw his weight around—we begin to see why his main target would be Cantor. Another difference lies in how his opponents dealt with this treatment. Schwarz, Weierstrass, and others were not happy about it but came through intact. Cantor, as we'll see later, had much more trouble in coming to terms with Kronecker, and with life's treatment in general.

Cantor and His Strange Ideas

Half a century ago, we might have read of another interesting reason for antipathy between the two men. Among the early 20th-century biographers of Cantor and Kronecker was Eric Temple Bell, an influential historian of mathematics with a silvery pen and a vivid imagination. In 1937, he wrote of their feud: "There is no more vicious academic hatred than that of one Jew for another when they disagree on purely scientific matters or when one is jealous or afraid of another."[9]

Aside from the questionable sociological aspect of the statement, Cantor—in spite of the Jewish-sounding name, and in spite of the fact that he chose the Hebrew letter aleph for his notation—was not Jewish. In fact, Bell writes elsewhere in the same article that "The family were Christians, the father having been converted to Protestantism; the mother was born a Roman Catholic."[10]

There may have been a Jewish element somewhere in Cantor's genealogical background, but he was born into a solidly religious Christian family and was later strongly attracted to, and became involved with, the Roman Catholic clergy. He even felt that set theory had been revealed to him by a Christian God. As he put it in 1896: "From me, Christian philosophy will be offered for the first time the true theory of the infinite."[11]

Another difference between Cantor and Kronecker lay in their cultural backgrounds. Kronecker's father was a businessman. Cantor's father was also in business, but the family was deeply steeped in the arts. As a youngster, Cantor showed talent in both music and drawing. Nevertheless, his mathematical ability and interest showed up by his mid-teens, and although his father wanted him to study engineering, he was able to overcome his father's objections and follow his own lead.

Cantor was born in 1845 in St. Petersburg and attended primary school there. But his father's health was not good and, when Georg was 11, they moved to Germany for its warmer climate. It appears that Georg never felt really comfortable in Germany and looked back fondly at his earlier years in Russia. In 1863, at the age of 18, he entered the University of Berlin and began his studies in earnest with

Weierstrass, Kummer, and Kronecker. He was also active in the Mathematical Society in Berlin and acted as president during 1864–1865.

Following a semester at the University of Göttingen in 1866, he wrote his doctoral thesis at the University of Berlin in 1867. Its title was "On Indeterminate Second-Degree Equations." In preparing to defend for his oral examination, however, he also dealt with the proposition "In Mathematics the Art of Asking Questions Is More Valuable than Solving Problems." He used as an example a question about number theory that Carl Friedrich Gauss had left open in his *Disquisitiones Arithmeticae* of 1801. This was an early hint of Cantor's special way of asking questions, which would later open up whole new areas of inquiry.

After receiving his doctorate in 1867, he taught briefly at a Berlin girl's school and then joined the faculty at the University of Halle. There he remained for his entire career: first as a lecturer (income from lecture fees only); then, in 1872, as an assistant professor; and, finally, as a full professor in 1879. This was a curious situation. He felt that he was being consigned to a basically second-rate school, shut off from inspiration and interaction with other high-caliber people, for his entire working life. Throughout his later career, he blamed Kronecker for being held down in this manner.

Yet throughout his career he *was* in touch with high-caliber people like Karl Weierstrass, Hermann A. Schwarz, Richard Dedekind, Gösta Mittag-Leffler, and Felix Klein. Further, being in Halle might be comparable to someone being at UMass Amherst when he really wanted to be at Harvard or MIT. Amherst is not second-rate, and being there doesn't cut one off from the rest of academia; but it isn't Harvard. In other words, although Cantor resented not being at Berlin or Göttingen, in truth Halle was really not as bad as he made it out to be. Finally, as we will see, he had powerful mood swings that may at times have intensified his unhappiness.

On the positive side, somehow he made it all work. He began to produce mathematical papers. The earliest had to do with the theory of numbers, reflecting the influences and the interests of Gauss and, ironically, Kronecker. Then Eduard Heine, one of the senior people at Halle, recognized something special in Cantor. Heine had wrestled with, and had done a paper on, an interesting question: if a function could be represented by a trigonometric series, was there only one such series? At Heine's suggestion, Cantor looked into the question and came up with an important proof of the uniqueness of such a series.

This was not a simple matter and was accomplished in several steps, each with a published paper showing extensions of his uniqueness theorem. Most of his early papers were published in the Swedish journal *Acta Mathematica*. This respected journal had been founded and was edited by the Swede Gösta Mittag-Leffler, who was among the first mathematicians to recognize Cantor's genius. Early on, Kronecker had made a suggestion to Cantor that proved very useful. Clearly, the two men were still on good terms at this point in their relationship.

Cantor continued to dig. He began thinking about the collection of numbers (or points), including irrational numbers, that would not conflict with a trigonometric representation, and in an 1872 paper, he defined irrational numbers in terms of convergent sequences of rational numbers. He was moving into territory that made Kronecker uneasy.

Cantor's uniqueness proof of a trigonometric series also involved the nature of point sets in a real line, so he began to explore and expand on the complexities of point sets and their connection with sets of other kinds of numbers.*

Infinite Set Theory

Philosophers, theologians, and mathematicians have wrestled with the concept of infinity, and its many implications, since the days of the Greeks. Galileo, for example, thought it important to point out in his classic *Dialogues Concerning Two New Sciences* (1638) that there are as many squares as there are natural numbers because, as he put it,

---

*The term *set* can have many meanings. In 1895, Cantor saw it as any gathering into a whole set M of distinct perceptual or mental objects *m*, which he called the elements of *M*.

"every square has its own root and every root its own square, while no square has more than one root and no root more than one square."[12]

In other words, if we consider the class, or set, of all positive integers, we have a huge, unbounded collection of numbers—a collection that, in itself, is beyond anyone's comprehension. Conceptually, however, it must exist. Then, considering Galileo's statement, we also have a set of squares, one for each of the positive integers. Galileo said there are as many squares as there are natural numbers. Yet Galileo knew there are also integers that are not squares: 2, 3, 5, 6, 7, and so on. How can there not be less squares than there are natural numbers? Galileo saw in this only a conundrum, a paradox, and went on to other things.

Cantor carried the idea further. He started with an idea put forth by Richard Dedekind, one of his early admirers. In 1872, Dedekind had defined a set as infinite if it could be put into one-to-one correspondence with one of its subsets. For example, if we list all the natural numbers, say $1, 2, 3 \ldots n \ldots$, we can easily place them in direct, one-to-one correspondence with their squares, $1, 4, 9 \ldots n^2$. .. Cantor took off from there. Both the full set $(1, 2, 3 \ldots n \ldots)$ and the subset $(1, 4, 9 \ldots n^2 \ldots)$, he said, are "countably infinite" or "denumerable."

Such "countably infinite" sets, he said, have the same "cardinality." To denote this level of cardinality, he used the first letter in the Hebrew alphabet, the aleph, along with the subscript zero, and termed it aleph-null ($\aleph_0$).

In other words, the set N of natural numbers can be put into a one-to-one correspondence with (has the same "power" as) a subset of itself. Therefore, the whole is equal to a part of itself. This, of course, is in direct opposition to Euclid's long-lasting axiom that the whole is greater than the part.

Now Cantor was really swinging. He began to create a new kind of arithmetic. He showed, for example, that if we add all the integers (aleph-null) to all the squared integers (also aleph-null), the result is still aleph-null!

$$\aleph_0 + \aleph_0 = \aleph_0$$

And, similarly,

$$\aleph_0 \times \aleph_0 = \aleph_0.$$

And so on. To some people, this might have seemed like nothing more than a game. Cantor saw that there was more to it, that it both required and was the beginning of a new kind of mathematics.

His choice of the aleph symbol was both clever and apt. He felt, mainly, that Greek and Roman letters were already widely used in mathematics and science, and that his mathematics deserved a unique symbol. But it was not until the early 1890s that he saw the need for a standardized symbol and formally introduced it. Prior to then he experimented with different notations. We use the aleph here to simplify the exposition.

Set Theory Is Born

The 1870s and 1880s were years of development in set theory. Dedekind had done some work with the one-to-one correspondence of infinite sets and subsets in 1872. Cantor went on to the next logical question: Are there sets that are uncountably infinite? That is, are there different sizes of infinities? He began this search by once again looking at his sets of numbers. He knew that the set of rational numbers could be put into one-to-one correspondence with the natural numbers, as could the algebraic numbers.*

Could this be done with the set of real numbers?[†] It took some doing, but by the end of December 1873, he could write to Dedekind that he had succeeded in proving that the set of real numbers could *not* be put into one-to-one correspondence with the set of natural numbers;[13] it was "uncountably infinite." Cantor used the name continuum for this set, and gave it the symbol c. At this point, set theory was born, and the concept of different sizes of infinity came into being.

---

*Rational numbers include integers and ratios of integers (fractions); algebraic numbers are the roots of polynomial equations with rational coefficients.

[†]Real numbers are defined as those that can be represented by decimals; also the set of numbers that includes all rational and irrational numbers.

With this proof, he had shown that the order of real numbers *was* higher than that of the natural numbers. He knew he had to publish this result. He also knew that there were mathematicians who viewed his work with strong reservations. In this paper, he would be dealing with both irrational numbers and sizes of infinity. He hoped to publish in *Crelle's Journal*. Unhappily, Kronecker, as an editor there, had the right to refuse any paper, and Kronecker had already signaled his unhappiness with the direction of Cantor's work. Furthermore, Kronecker's opinions, including his feelings about irrational numbers, were widely known in the mathematical world; if other mathematicians saw a paper spelling out Cantor's new results, they might very well also object, if only to please Kronecker.

Cantor decided on a clever ruse. He figured that many of the people in the field, perhaps including Kronecker, might very well just scan the title of the work to see if there was anything objectionable in it. He titled his paper, "On a Property of the Collection of All Real Algebraic Numbers." It therefore appeared from the title to be simply a new proof of an earlier theorem of Joseph Liouville's that non-algebraic real numbers do exist. Superficially, it appeared that he was writing only about algebraic numbers. The ploy worked. The paper slipped through and was published in *Crelle's Journal* in 1874. Set theory was launched—but it had to be buried in a paper that seemed to be on another subject.

Now, however, the fat was in the fire. From then on, Cantor knew, Kronecker would be more careful.

Conflict Begins

Things were quiet for a while, and Cantor moved on to his next challenge. In 1877, he found a property of infinity that even he was thrown by. In a letter to Dedekind in 1874, he had posed the following question: "Can a surface (say, a square that includes the boundaries) be uniquely referred to a line (say, a straight-line segment that includes the end points) so that for every point of the surface there is a corresponding point of the line and conversely, for every point of the line there is a corresponding point of the surface? Methinks that answering this question would be no easy job, despite the fact that the answer seems so clearly to be 'no' that proof appears almost unnecessary."[14]

The first remarkable part of this story is that Cantor should pose such a question. The second is that, three years later, he could write again to Dedekind that he had shown that the answer is "yes." Generalizing somewhat, he said that continuous spaces of n dimensions *could* be put into one-to-one correspondence with (had the same "power" as) the set of points on a line. "I see it," he wrote, "but I don't believe it."[15] His proof was somewhat unwieldy but quite correct.

Dedekind congratulated Cantor on his new finding but warned him that publication would be difficult. He was right. Cantor sent off his paper with this finding to *Crelle's Journal* on July 12, 1877. In spite of the controversial findings—that all continuous lines, planes, or surfaces had the same order of infinity—all seemed well when the editor promised to publish it and Weierstrass promised to promote it when it appeared.

Yet as time passed, it was clear that no steps were being taken toward its publication. Cantor suspected that Kronecker was acting behind the scenes to prevent its appearance. He became increasingly agitated and wrote to Dedekind that he was thinking of withdrawing the paper and trying to publish it elsewhere—even though *Crelle's Journal* had published his earlier work. Dedekind persuaded him to hold on for a bit, and perhaps exerted some influence. In any case, this important paper, sometimes referred to as Cantor's *Beitrag*, did finally appear the following year.

Though it was published, Cantor was upset about the long delay and the fact that Kronecker could exert his wiles even to that extent.

It's important to understand that part of the driving force behind Kronecker's obstructive activities was a purely mathematical one, an honest disagreement with Cantor's mathematical ideas. Kronecker felt that Cantor was playing with inconsequential concepts and that he should not be allowed to publish ideas that would go nowhere. This would not be the first time Kronecker had done this, having acted as interference against other papers that dealt with irrational numbers and infinity. Furthermore, Cantor's proof was based on one-to-one mappings between irrational and real numbers, and as far as Kronecker was concerned, irrational numbers do not exist.

Cantor's paper did finally appear, though, and being published in *Crelle's Journal* brought him into highly respected territory. While by no means everyone agreed with him, it was clear that he was establishing himself as a revolutionary thinker. If his feelings and beliefs were right, he was creating a whole new mathematics, of which he was the obvious master. He began to feel increasingly lonely, and unhappy about his apparent inability to move up in the world. He wished desperately to join the faculty at a major German university—perhaps Göttingen or, preferably, the University of Berlin. He also felt he was being grossly underpaid in comparison with stars like Schwarz, L. Fuchs, and, especially, Kronecker. In fact, he had applied to the University of Berlin and was told by Weierstrass that the reason his application was turned down was Kronecker's large salary.

Cantor was unhappy about this but suspected there was more to the story. And he was right. Kronecker, seeing that he couldn't actually prevent Cantor from publishing, put some of his efforts into attacks on Cantor and his work. Kronecker's technique included suggesting—though never in print!—that not only was Cantor's work humbug, but that he was a charlatan and a corrupter of youth, that he was luring these young people into a "dangerous world of mathematical insanity."[16] In any case, Kronecker's position of power made such epithets powerful weapons in his subtle war.

According to Morris Kline, a noted historian of mathematics, Kronecker's attacks did indeed leave mathematicians suspicious of Cantor's work[17] and may well have been a factor in Cantor's not being invited to join the faculty at a more prestigious school. Cantor spent his entire career at Halle. It's worth noting, however, that the Halle authorities were very generous in their treatment of Cantor, as when they arranged for leave from his teaching duties in his coming times of need.

Like Kronecker, Cantor saw their conflict as at least partly a battle over whose view of mathematics was the correct one, and even in this ostensibly unemotional arena, he believed that the outcome would not be decided in a pure search for truth. Rather, he saw it as "a question of power, and that kind of question can never be decided by way of persuasion; the question is which ideas are the most powerful, comprehensive, and fruitful, Kronecker's or mine; only success will in time decide our struggle!"[18]

In the meantime, Cantor was not sitting idly by while Kronecker was carrying out his attacks. After his application to Berlin was turned down, for example, Cantor wrote a letter directly to the minister of education in Berlin, complaining about these attacks. Then, early in 1884, he wrote to Mittag-Leffler, "I never thought . . . that I would actually come to Berlin. But since I plan to do so eventually and since I know that for years Schwarz and Kronecker have intrigued terribly against me, in fear that one day I would come to Berlin, I regarded it as my duty to take the initiative and turn to the Minister myself. I knew precisely the immediate effect this would have: that in fact Kronecker would flare up as if stung by a scorpion, and with his reserve troops would strike up such a howl that Berlin would think it had been transported to the sandy deserts of Africa, with its lions, tigers, and hyenas. It seems I have actually achieved this goal."[19]

Cantor, of course, was right; Kronecker did respond. Yet if Cantor thought there would be a howl, there he was wrong. Once again, with a subtle move rather than a howl of anger, Kronecker acted behind the scenes. He may have had an inkling that there was some mental instability lurking in Cantor and, it would seem, wondered what he could do to bring on something serious. In January 1884, Kronecker contacted Mittag-Leffler and suggested doing a short article in *Acta Mathematica*, in which he would show "that the results of modern function theory and set theory are of no real significance."[20]

When Cantor got wind of this, it seemed clear to him that Kronecker was trying to deprive him of the one major publishing outlet with a sympathetic editor that was still open to him, just as Kronecker had tried to do with Cantor's earlier article in *Crelle's Journal*. Seeing this as a polemical article, Cantor threatened to stop sending his own work if Mittag-Leffler should accept such an article. With his own importance in the field still growing, this threat had some bite. Yet Kronecker never sent anything; the likelihood is that he was just trying to goad Cantor into some sort of uneven or unpleasant behavior. As it turned out, Cantor did later act on his threat—he cut off all contact with the journal—though for another reason. Still, Cantor's response, as Kronecker hoped, undoubtedly did some damage to Cantor's relationship with an important supporter.

Cantor's Later Years

None of this, however, stopped the Cantor production line from continuing to roll out its new mathematics. In 1879, Cantor began some new work and developed it in a series of six papers published from 1879 to 1884. Cantor was back at work on a problem that had been haunting him. He had earlier found sets that were infinitely countable (aleph-null) and sets that were infinitely uncountable, for example, the real numbers, which he had tentatively named c. But, he wondered, are there sets that are intermediate in size between aleph-null and c?

He believed, and hoped to prove, that the answer was no—that aleph-null and c behave in his arithmetic like 0 and 1 behave among the integers. The latter are the first two integers in our number system, and no other whole numbers fit between them. If a similar situation held in set theory, then c would be equivalent to aleph-one, which he defined as the next order (or level) of infinity after aleph-null, and he could build up his aleph-based number system from there.

There are several ways of stating what came to be called his continuum hypothesis. Simply stated:

$$\aleph_1 = c.$$

He felt this was true but had to prove it.

In the fifth of the six papers, he showed his mixed feelings about his position in the mathematical world. He wrote, "Daring as this [work] might appear, I express not only the hope but also the firm conviction that in due course this generalization will be acknowledged as a quite simple, appropriate, and natural step. Still I am well aware that by adopting such a procedure I am putting myself in opposition to widespread views regarding infinity in mathematics and to current opinions on the nature of number."[21]

He wrestled almost continually with the continuum hypothesis over the balance of his life and career. In the course of this work, he repeatedly wrote to Mittag-Leffler saying he had the answer; then he wrote again, retracting that proof, and claimed that the continuum hypothesis was not true. And so it went. He kept trying, with the problem becoming an obsession with him. In the sixth paper, he wrote that the answer would be forthcoming.

It would seem that there must have been something lacking in his mathematical skills, but that was not the case. What he didn't know was that he was trying to solve a problem that had no answer. That is, as we'll see in a later chapter, the continuum hypothesis had no answer in the mathematical system as it stood in Cantor's day.

Yet the maddening process of seeing the answer one way one day and another the next cost him dearly. His obsession with the problem, plus his reaction to what he saw as the unfair and even vindictive treatment by his colleagues, especially Kronecker, had a severe effect on his mind.

The Point of Madness

Earlier biographies of Cantor's life saw these actions—his obsession with the continuum hypothesis and Kronecker's assaults—as factors that drove Cantor to the point of madness. There were other setbacks. In 1881, an opening occurred at Halle, and Cantor managed to arrange for Dedekind, who was teaching in a secondary school, to be offered Heine's chair. Cantor was obviously looking for both friendship and some academic stimulation; but, in addition, had he been able to bring Dedekind in, Halle itself would have become a more important center of mathematics. Early in 1882, Dedekind declined, no doubt reflecting some fear of a clash of personalities. Cantor's next two recommended names also declined. A new list was drawn up, but the new man and Cantor never formed a close relationship.

Cantor did indeed end up spending various periods in a mental institution. By May of 1884, at age 41, he suffered the first of a number of serious mental breakdowns that were to plague him over the next 33 years of his life. He entered the Halle Nervenklinik, a university-based mental clinic, for treatment, which lasted for somewhat longer than a month.

Again, we turn to Bell for a fascinating scenario. Writing of the young Cantor, he stated, "Georg was determined to become a mathematician, but his practical father . . . obstinately tried to force him into engineering as a more promising bread-and-butter profession. . . . Loving his father devotedly and being of a deeply religious nature, young Cantor could not see that the old man was merely rationalizing his own greed for money. Thus began the first warping of Georg Cantor's acutely sensitive mind." Also, "In the process of trying to please his father against the promptings of his own instincts Georg Cantor sowed the seeds of the self-distrust which was to make him an easy victim for Kronecker's vicious attack in later life and cause him to doubt the value of his work. . . .The father gave in when the mischief was already done."[22] A nicely argued scenario, but basically untrue. Georg's father appears to have been a solid, understanding parent who was genuinely interested in both Georg and his career.

Later writers feel that Cantor was actually a victim of a bipolar disorder (manic/depressive illness), and that his mathematical setbacks and difficult personal relationships, especially with Kronecker, may well have been magnified by his mental condition but were not its cause. As supporting evidence, we also have the fact that he at times exhibited "doubtless exaggerated feelings of persecution which he felt upon himself from his [university] colleagues."[23]

Upon emerging from the Nervenklinik, after this first stay, something led him to think that he could still work things out with Kronecker, and upon his release, he actually contacted Kronecker and attempted a reconciliation. Kronecker responded without rancor, and Cantor tried in a second letter to explain some further details of new work he was doing. In his letter he wrote, as soothingly as he could, "I am of the opinion that the greatest part of what I have done scientifically in the last few years, which I include under the rubric of set theory, is not so very much opposed to the demands which you place upon 'concrete' mathematics as you seem to believe. It may be the fault of the presentation (which may not be entirely clear), that you have given less attention to the concrete mathematics in my research than to its other, namely, philosophical content."[24]

They even met and spent an evening together in October 1884, but the basic chasm between them was just too large, and little was accomplished. Cantor later described the meeting in a letter to Mittag-Leffler: "It seems to me of no small account that he [Kronecker] and his preconceptions have been turned from the offensive to the defensive by the success of my work. As he told me, he wants to publish soon his opinions concerning arithmetic and the theory of functions. I wish it luck!"[25]

What were Kronecker's real feelings about Cantor? In an 1885 letter from Sophie Kowalevsky to Mittag-Leffler, she writes that Kronecker "was quite bitter about Cantor."[26]

In 1885, Cantor took another hit, this time from his own friend and supporter Mittag-Leffler. Early in the year, Cantor sent off two letters for publication in the *Acta Mathematica*, spelling out some new work. In March, Mittag-Leffler wrote back, saying that since it didn't contain the proof of any significant result, publication of this work, before Cantor had been able to explain these results, would damage his reputation. He felt the publications were about a century too early!–that no one would understand them if published now, and that the theory behind it would be rediscovered by someone else a hundred years in the future and credited to him.

Cantor was devastated by this turn of events. Mittag-Leffler was trying to do the right thing, but it shows that even he didn't fully understand the importance and the power of Cantor's work. Cantor felt that he had been abandoned by the only important mathematician who had supported his crusade. Here, then, was the conclusion to the drama begun earlier by Kronecker. Cantor asked for all of his papers back from *Acta Mathematica* and never published there again. Mittag-Leffler had not abandoned Cantor, but to all intents and purposes, that was the result.

Cantor, reacting perhaps too strongly, saw his future in mathematics dimming perceptibly–indeed, looking hopeless. He began to devote more of his time to other pursuits, including philosophy and theology. He also found a greater level of support for his mathematical work among theologians of the Roman Catholic

Church than he had seen among mathematicians.[27]

At the same time, the periods between his mental attacks were getting shorter, and he was spending more time at the Klinik. The death of his mother in 1896 and of his younger brother and youngest son in 1899 added more emotional turmoil to his life.

Invitations to the Dance

Unable to crack the challenge of the continuum hypothesis, and never really reconciled to his position in the mathematical world, Cantor might at times have used his periods of illness as a respite from the unpleasantnesses and even agonies of his personal and mathematical disappointments. At various periods during these years, he flung himself into efforts to show that the British philosopher-statesman Francis Bacon had authored the Shakespearean plays. One writer, Nathalie Charraud, goes so far as to suggest that Cantor's attempts to expose the true Shakespeare to the world were in some way allied with his wish to expose the true Kronecker to the world.[28]

His mind also turned toward religion. In these periods, he raised the status of the continuum hypothesis to dogma—and thus beyond the need for proof—and declared that "from me, Christian philosophy will be offered for the first time the true theory of the infinite."[29]

Between incarcerations, however, his active mind kept going. In fact, the role of Cantor's mental illness was almost certainly not entirely a negative one. One of the common characteristics of bipolar illness is a kind of hypomania, or exuberance, in the "up" phases that—in some victims—generates bursts of creativity and focused work. This could very well have been the case with Cantor.

In general, though, in his later years he concentrated more on the philosophical aspects of his work and was able to engage in discussions of it with a variety of scholars. He had also long had an interest in encouraging promising young people and in the earlier years was able to bring to life his dream of founding the Deutsche Mathematiker-Vereinigung (DMV, or German Mathematicians' Union), an official mathematical society. He felt, first, that it would provide a forum for these young mathematicians, so that they might not have to go through what he had been subjected to. In addition, he believed that young people might be better able to understand and deal with his kind of mathematics.

Cantor chaired the first meeting, which was held at his university in Halle in September 1891, and he served as its president until 1893. He was also going to present a paper at that first meeting–indeed, his first paper on new mathematical research in more than five years. He was showing a new proof for the existence of nondenumerable sets. In the paper, he introduced a means of showing that given any set, the set of all its subsets was always of a power greater than that of the parent set itself. Central to this work was a method that came to be well known as his diagonalization method. Although the diagonalization method was a solid piece of work, his comparison of the powers of sets and subsets would come back to haunt him later on.

In spite of the now open antagonism between himself and Kronecker, he nevertheless invited Kronecker to address this inaugural meeting. There were, of course, obvious reasons for the invitation–mainly, Kronecker's position and high reputation. One could hardly ignore him. Yet there was also a deeper, more subtle purpose. Kronecker's attacks had always been behind the scenes and had always been made in discussions and lectures, certainly never in print. Cantor hoped that in such a forum, Kronecker would almost be forced to speak out, in public, his real feelings about set theory, and that this would reveal his bias to the mathematical world.

As Cantor had done 13 years earlier in the *Crelle's Journal* article, in order to slip past Kronecker's radar, he tailored the title of his own paper in a special way. This time he titled it, "On an Elementary Property of Set Theory." He felt that irrational numbers remained controversial, but that with his new method he need no longer rely on them or even on the general idea of infinite sets.

Whether Kronecker would have been lured into this trap, we'll never know. Kronecker's wife was injured in a mountain climbing accident, and he sent word that he wished the meeting success, but that he could not appear.

Cantor later confided to Mittag-Leffler that it was probably just as well that Kronecker did not appear, for it would have provided him with an excellent opportunity to malign Cantor behind the scenes and at Cantor's own university. As it turned out, Kronecker himself died in December of the same year.

The meeting was rated a success, and Cantor's paper was later published in the first volume of the *Jahresbericht* of the DMV. Cantor had shown that given any set, the set of all its subsets was always of a higher power (had a higher cardinal number than) the parent set itself. In the paper, he placed new emphasis on the finite and the infinite cardinal numbers, which were shown to be just the finite and the infinite powers of sets, respectively.

In the meantime, Cantor's feelings about Germany and German mathematicians were, not surprisingly, increasingly negative, and so it makes sense that he shortly became deeply involved in the planning for the First International Congress of Mathematicians, to be convened in Zurich in 1897. As with the DMV, he hoped to create a more encouraging forum for new ideas.

Possibly feeling free for the first time of the baleful eye of Kronecker, Cantor produced an updated, detailed presentation of his transfinite set theory. It would be presented in two articles in the *Mathematische Annalen* in 1895 and 1897.

A New Century

In 1888, Cantor had written, "My theory stands firm as a rock; every arrow directed against it will return quickly to its archer. How do I know this? Because I have studied it from all sides for many years; because I have examined all objections which have ever been made against the infinite numbers."[30]

For a while, it began to seem that everything in mathematics was going to be defined or explainable in terms of sets—in fact, that set theory would eventually become the foundation of mathematics.

By the turn of the century, however, Cantor was feeling much less confident. First, certain paradoxes had been discovered in his work, by him as well as others, that caused him and his followers considerable anguish. A paradox in this case is a statement that derives contradictory conclusions from acceptable premises. A simple example is the famous story of the barber of Seville. He claims to shave all the men in the city of Seville who do not shave themselves. Does he shave himself?

The best-known paradox in set theory was put forth in 1901 by Bertrand A. W. Russell, the widely known British philosopher turned mathematician. Russell asked what would seem at first to be a simple question, yet it shook the very foundations of set theory and all it stood for in the larger world of mathematics. Russell considered the fact that virtually anything can be grouped into sets, which is what makes the concept so powerful. He postulated the set of all sets that are not members of themselves, and he called this set R. Then he asked, is R a member of itself? As with the barber, if R is a member of itself, it isn't; if it is not, it is.

Russell's paradox was not the first paradox, or the only one, that had been found in set theory. In fact, Cantor's friend Ernst Zermelo had earlier come up with something similar but had not thought it worth pursuing or publishing.

Yet the implications of the paradoxes were powerful. Early set theory had considered the possibility of a universal set, a set that contained everything. This was now seen to be impossible. Not everything could be a set. We begin to see what a serious problem these paradoxes were to mathematicians and logicians alike. Like the other paradoxes, Russell's paradox hung around like a big elephant in the garden.

To some mathematicians, however, none of the various negative developments, including the paradoxes, seemed to overturn Cantor's basic results in transfinite arithmetic. In fact, Cantor's new ideas and theories were being recognized as important in the further development of analysis, function theory, topology, and non-Euclidean geometry, and indeed as the basis for a more fundamental understanding of mathematics in general. (Today, elementary set theory is likely to be taught in various high school mathematics courses, and especially in probability and topology.)

In the early years of the 20th century, Cantor was still communicating with some of his colleagues and actively defending his work. Happily, his work gained wide acceptance while he was still alive, and the organizers of conferences and award presentations would have been happy to invite him. Unhappily, though, his illness had by then progressed to the point where he was really in no condition to participate in public meetings. He took a leave from teaching in the summer of 1899 and the winter terms of 1902–1903 and 1904–1905, when he spent some time in sanitoria. After this, he spent increasing periods in the Halle clinic, for example, from October 22, 1907, until June 15, 1908, and from September 28, 1911, to June 18, 1912, when he was moved to a different sanitorium.

Cantor was admitted to the Halle clinic for the last time on May 11, 1917. He was not happy about going and wrote repeatedly to his wife asking to come home, without results. World War I was in full swing at the time, and food was scarce, which did not help matters. On January 6, 1918, this extraordinary man, still at the clinic, died of a heart attack at the age of 73.

In a small area in the center of Halle, a plaque is mounted that shows Cantor's face, some numbers suggesting his diagonalization proof, and a sentence in German that states, "The essence of mathematics lies precisely in its freedom." Cantor preferred to use the expression *free mathematics* rather than the more common term *pure mathematics*.

Summation

For many years, writers touching on the relationship between Cantor and Kronecker saw Kronecker as "both wrong and unfair." Their information almost invariably came from Bell. As he put it in his somewhat flowery but always compelling manner: "Seeing mathematics headed for the madhouse under Cantor's leadership, and being passionately devoted to what he considered the truth of mathematics, Kronecker attacked 'the positive theory of infinity' and its hypersensitive author vigorously and viciously with every weapon that came to his hand, and the tragic outcome was that not the theory of sets went to the asylum, but Cantor. Kronecker's attack broke the creator of the theory."[31]

Later on, Bell softened this a bit: "Kronecker perhaps has been blamed too severely for Cantor's tragedy; his attack was one of many contributing causes."[32] Yet the implication remains the same: Kronecker is the offender, the bad guy; many of the writers who came after Bell took the same approach.

We have already seen that Kronecker's activities may well have exacerbated Cantor's illnesses but were most likely not the cause. As to just how "vigorously and viciously" he attacked Cantor, there is now some question about that, too.

Harold M. Edwards, a professor of mathematics at New York University, has studied both Kronecker's life and work, and he says, in essence, "Kronecker is so often depicted as dogmatic, extreme and vitriolic. . . . I believe in the reasonable Kronecker and not the vitriolic Kronecker."[33] He explains, "Whereas Kronecker played an enormous role in Cantor's distinctly paranoid world view, I doubt that Cantor played a very large role in Kronecker's. To Kronecker, I suspect, Cantor was simply another young man who had followed Weierstrass onto the wrong path and whose formulations of mathematical ideas were hopelessly misguided."[34]

Edwards points out that "One finds reverential references to Kronecker's works for 10 or 15 years after his death, but, after that . . . with rare exceptions, the mathematicians who knew Kronecker's work knew it second hand." He argues that this came about in the following way:

> The strength of the Weierstrass school, and their bad feelings toward Kronecker (as evidenced by Mittag-Leffler's address in 1900), surely played a role. Another figure who played a role was Dedekind, who exercised a major influence both through his own works and through the effect he had on such younger mathematicians as Weber, Cantor, and Hilbert. Dedekind created a style of mathematics and an attitude toward the foundations of mathematics that tended to make Kronecker's works look much more difficult than they were. . . . And, lastly, Hilbert and Cantor each in his own right did much to turn the younger generation away from Kronecker.[35]

As an example of such a result, John D. Barrow writes in his well-received book *Pi in the Sky* that Cantor's "name is remembered where Kronecker's is now largely forgotten."[36] A mathematician whose works are no longer read is more easily seen as the "bad guy."

Morris Kline points out, "Kronecker had no supporters of his philosophy in his day and for almost twenty-five years no one pursued his ideas."[37] After the paradoxes were discovered, however (around the turn of the century), some of Kronecker's work was picked up and further developed by mathematicians like Poincaré and Brouwer, whom we discuss in later chapters.

Edwards writes, "I believe that Kronecker's best hope of survival comes from . . . the tendency, fostered by the advent of computers, toward algorithmic thinking. . . . It is my hope," he adds, "that the revival of this long-dormant point of view will bring with it a renewed appreciation of his legacy"[38] and perhaps, another look at Kronecker himself. Joseph Dauben appears to agree with Edwards, at least in principle, that Kronecker has been unfairly stigmatized.[39]

Randall Collins and Sal Restivo, two sociologists who have studied this feud, make an interesting point: "The struggle between Kronecker and Cantor . . . was not a conflict between traditional and innovative forms of mathematics, but between rival new paradigms. Kronecker was not a mathematical traditionalist; in opposing an actual infinity and irrational, transcendental, and transfinite numbers, he was forced to reconstruct mathematics on a radically new basis. He foreshadowed the 20th-century school of intuitionists, just as Cantor pioneered in what became the formalist program. Both sides pressed for greater rigor in mathematics, but were divided sharply on how to achieve it."[40]

Regardless of whether Kronecker was actually and directly at fault for Cantor's unhappy state of affairs, there were some positive results from the Kronecker-Cantor conflict. Cantor's initial ideas were not solidly grounded—for example, they were based more on ideas than on axioms—which is not surprising considering their newness and originality. The knowledge that critics, especially Kronecker, were waiting to pounce forced him and several followers to continue to dig and eventually to produce a more solid foundation for his theory.

Even so, when his set theory was finally launched on a broader scale, it came close to being wrecked by a series of paradoxes that arose as a result. The German mathematician Ernst Zermelo, one of his followers, went on to axiomatize set theory in an attempt to save it. Yet he, too, would shortly find himself embroiled in a bitter debate about the value of his work.

7

Borel versus Zermelo

The "Notorious Axiom"

The Second International Congress of Mathematicians was held at the Paris World Exposition of 1900. There the well-known German mathematician David Hilbert delivered an address in which he listed the major unsolved problems in mathematics at that time. He had come up with a total of 23 problems and mentioned 10 in his speech. Number one on his list was the still unfound proof for Georg Cantor's continuum hypothesis.

Cantor, you will recall from the last chapter, had set up a system of transfinite cardinal numbers, which he had put into an ordered arrangement of alephs ($\aleph_0$, $\aleph_1$, $\aleph_2$ . . .). There were, he believed, no cardinal numbers outside this system of alephs. Yet before Cantor could prove that each cardinal number could be placed within this system, he had to compare every possible pair of set constituents via his one-to-one correspondence method. Furthermore, the infinite cardinals had to show the same ordering principle as was the case with real numbers in a line. Thus, for any two real numbers, they had to be equal ($a = b$), or one had to be greater than the other: ($b > a$) or ($b < a$).

For this to work, Cantor had to set up a specific property, which he called the well-ordering principle. A set is classified as well ordered if it automatically has a smallest element. Thus the set of all positive integers in their natural order is well ordered because it begins with a first or smallest element, namely, the number 1. On the other hand, the set of all integers—which includes negative numbers—is not well ordered because we would have to tinker with it first to establish a first or smallest element. Therefore, the set of positive integers and the set of integers would have the same cardinal number but would be of different order types.

Cantor felt he was on to something important. In a lengthy article (the *Grundlagen*, 1883) he wrote, "The concept of *well-ordered sets* turns out to be essential to the entire theory of point-sets. It is always possible to bring any *well-defined* set into the *form* of a *well-ordered* set. Since this law of thought appears to me to be fundamental, rich in consequences, and particularly marvelous for its general validity, I shall return to it in a later article."[1]

If he could prove the well-ordering principle, it would permit him to show that each transfinite cardinal number was equivalent to one of his alephs, an important step toward proving the continuum hypothesis. More specifically, he was trying, at least in his more lucid periods, to prove that aleph-one (which he defined as the next order of infinity following aleph-null) equaled c, the power of the continuum or set of all real numbers ($\aleph_1 = c$). That would show, he said, that there is no intermediate form of infinity, no set of elements whose power is greater than that of the set of all natural numbers ($\aleph_0$) but less than that of the set of all real numbers (c).

He couldn't do it. The proof continued to elude him.

Yet his set theory was stirring the mathematical world, both positively and negatively. In the meantime, Cantor spent some more time in the Halle Nervenklinik. Then, in 1903, he was back at work at his mathematics, and he spoke at a meeting of the Deutsche

Mathematiker-Vereinigung, answering some questions raised earlier by French mathematicians. A year later, he was awarded one of mathematics' highest honors—in fact, the highest that England's Royal Society can confer, the Sylvester Medal.

In the same year, however, Cantor found himself face-to-face with what looked like a major challenge to his theory. Jules König, a recognized mathematician from Budapest, read a paper at the Third International Congress held at Heidelberg (in 1904) that claimed that the power of Cantor's continuum was not any aleph, let alone aleph-one.

Today, a report on pure mathematics, even one made at a major congress of mathematicians, might not be reported in the public press. In that day, however, König's report made page one news. We can only guess at Cantor's reaction. We know he refused to accept the proof but could find no mistakes or gaps in König's reasoning. Furthermore, König had a good reputation among his peers.

Less than a day later, however, a young mathematician from the University of Göttingen, Ernst Zermelo, came to Cantor's rescue. Zermelo showed that one of König's premises had been faulty, and so his proof was not a solid one, but Cantor knew that the respite was only temporary. Until he or someone could prove the continuum hypothesis, could prove that the continuum was indeed an aleph, his entire corpus of work remained little more than a theoretical construct.

Happily, Cantor had his defenders as well as his detractors.

Zermelo

Ernst Friedrich Ferdinand Zermelo, born in 1871, grew up in Berlin. He studied mathematics, physics, and philosophy at universities in Berlin, Halle, and Freiburg and received instruction from some eminent teachers, including Max Planck, Edmund Husserl, and Herman A. Schwarz. He wrote a dissertation on the calculus of variations, which extended a method of Karl Weierstrass's, and received his doctorate from the University of Berlin in 1894. The University of

Göttingen appointed him privatdozent in 1899. By the winter semester of 1900–1901, he had become interested in and was lecturing on set theory, and at the Third International Congress in 1904, as we saw earlier, it was he who saved the day for Cantor by showing the weakness in König's attack.

Yet Zermelo, like Cantor, continued to fear that set theory was susceptible to more attacks in its existing form. In the early days of set theory, for example, Cantor had used rather haphazard methods in deciding what elements could go into making up sets. He also stated that every well-defined set can be brought into the form of a well-ordered set, but he never carried this further. The first step, Zermelo felt, would be to prove Cantor's well-ordering principle.

Zermelo supplied the critical item needed for proving the well-ordering principle. He maintained that from any given collection of nonempty sets, one could choose exactly one element from each set and thereby form a new set. In other words, given any collection of sets, there exists a method of designating a particular element of each set as a special element of that set. Thus, every set can be well ordered if it is assumed that in each nonempty subset, one element can be chosen, or designated, as a special element. This assumption is called Zermelo's axiom of choice.

The axiom of choice rang a bell with many mathematicians; mathematics needed it, they felt, and it simplified many proofs. The idea involved making infinitely many choices (conceptually), an idea that was not entirely new, having been played with by Cantor and other mathematicians earlier. In fact, claimed Zermelo, "it is applied without hesitation everywhere in mathematical deduction."[2] Yet Zermelo's construction was the first solid statement of the idea, and it stuck. It brought fame to Zermelo, and he was appointed a titular (titled) professor at Göttingen in 1905.

His axiom of choice, however, also set off a storm of controversy. Eric Temple Bell came to regard it as the "notorious axiom."[3] The controversy—pro and, mostly, con—erupted in many countries, including Germany, England, Hungary, Holland, Italy, and the United States,[4] with much of it centered among French mathematicians. Foremost among the objectors was Émile Borel.

Borel

Émile Félix-Édouard-Justin Borel was born in Saint-Affrique, Avey-ron, France, in 1871, the same year in which Zermelo was born. His mathematical talent also showed early. Known as a prodigy by age 11, he left his local school for study at the nearby Montauban lycée. At age 19, he entered the École Polytechnique and published two papers there in his first year. First in his class of 1893, he was promptly invited to join the faculty at the University of Lille. By 1894, at age 23, he had his D.Sc. from the École Normale Supérieure, where he quickly began building a solid reputation. By 1911, he was its scientific director.

In 1901, he married, and his interests began to broaden to appli-cations of mathematics and to public affairs as well. Yet this seemed not to interfere with his theoretical mathematical output and interests. Among these was a special interest in set theory, and in 1898, Borel published a critical analysis of Cantorian set theory in his *Leçons sur la théorie des fonctions*. So when Zermelo's proof of the axiom of choice appeared in *Mathematische Annalen* in 1904, the editors of the follow-ing issue included an international sampling of comment and criti-cism, but especially that of Émile Borel, from whom the editors knew they could expect some lively comments.

For example, as Borel put it at the end of his argument: "It seems to me that the objection against it [Zermelo's proof] is also valid for every reasoning where one assumes an arbitrary choice made an uncountable number of times, for such reasoning does not belong in mathematics."[5]

Russell, trying to clarify the situation, gave this example in 1905:

> Given alpha-null pairs of boots, let it be required to prove that the number of boots is even. This will be the case if all the boots can be divided into two classes which are mutually similar. If now each pair has the right and left boots different, we need only put all the right boots in one class, and all the left boots in another: the class of right boots is similar to the class of left boots, and our problem is solved. But, if the right and left boots in each pair are

indistinguishable, we cannot discover any property belonging to exactly half the boots. Hence we cannot divide the boots into two equal parts, and we cannot prove that the number of them is even. If the number of pairs were finite, we could simply choose one out of each pair; but we cannot choose one out of each of an infinite number of pairs unless we have a rule of choice, and in the present case no rule can be found.[6]

The controversy over the axiom of choice had some similarity with another well-known axiom, Euclid's parallel postulate and the questions that arose with the advent of non-Euclidean geometry. This time, the controversy circled around the question of what is an admissible method in mathematics. Zermelo's method provided no constructive definition of either elements or methods involved in its use. Borel was resolutely opposed to nonconstructive methods.

In essence, Borel was directly challenging Zermelo's claim that from each nonempty subset, one element can be chosen, or designated, as a special element, and that one could thereby create a well-ordered set. Borel and his group also argued against the axiom of choice because it calls for an infinity of operations, which is impossible to conceive.

Borel agreed, however, that Zermelo had been trying to solve an important problem, and his strong objections generated considerable controversy. He then gathered together the opinions of several leading French mathematicians on the subject—J. Hadamard, René Baire, and H. Lebesgue, plus his own—and published the results in the *Bulletin de la Société Mathématique* in 1905 under the title "Cinq lettres sur la théorie des ensembles." He was fairly even-handed, though: Hadamard supported Zermelo; Baire and Lebesgue stood on his side. Hadamard complained that Borel and his group were demanding from Zermelo something well beyond anything he had claimed or even wanted.

The interchanges, however, went in all directions. Baire, for example, wrote in a 1905 letter to Hadamard, "Borel has communicated to me the letter in which you express your viewpoint in the great debate resulting from Zermelo's note. . . . As you know, I share

Borel's opinion in general, and if I depart from it, it is to go further than he does."

Later, in the same letter, Baire wrote, "Zermelo says: 'Let us suppose that to each subset of M there corresponds one of its elements.' This supposition is, I grant, in no way contradictory. Hence all that it proves, as far as I am concerned, is that we do not perceive a contradiction in supposing that, in each set which is defined for us, the elements are positionally related to each other in exactly the same way as the elements of a well-ordered set. In order to say, then, that one has established that every set can be put in the form of a well-ordered set, the meaning of these words must be extended in an extraordinary way and, I would add, a fallacious one."[7]

A few years later, in 1912, Borel summarized his version of the argument. In the following writeup, Borel begins by challenging a method created by Cantor that involves use of successive decimal numbers to prove that the size of the set of all real numbers is greater than the size of the set of all integers and all rational numbers, and thus that there are different orders of infinity.

Borel wrote:

It is possible to define a bounded decimal number by demanding that a thousand persons each write an arbitrary digit. One will have a well-defined number if the persons are put in line each writing in turn a digit at the end of the digits already written by those in front in the line. The disagreement starts when one tries to extend this procedure to an unbounded decimal number. I do not suppose that people dream of actually having an infinite number of persons each writing an arbitrary digit, but I believe that Mr. Zermelo and Mr. Hadamard think that it is possible to regard such a choice realized in a perfectly well-defined way even if the complete definition of the number contains an infinite number of words. For my part I think it is possible to pose problems about probability for decimal numbers which are obtained by choosing the digits either randomly or by imposing certain restrictions on the choice-restrictions leaving some randomness to the choice. But I think it is impossible to talk about one of these numbers for the reason that if one denotes it by A, two mathematicians talking about *A* would never be sure whether they were talking about the same number.[8]

Yet another objection was Cantor's use of the "set of all sets," which had led initially to the paradoxes we considered briefly in chapter 6. Borel's group argued that the concept had not been properly defined.

The continuum, too, seemed to stick in the throats of members of the Borel group. Baire, for example, refused to believe that the continuum could ever be well-ordered; the two concepts, he felt, were just too different; he saw the continuum as the collection of all infinite sequences of integers.

Axiomatics

The controversy was never fully resolved, but the pro-Cantor side dealt with it in several different ways. One route went by way of the paradoxes.

Hilbert had been fascinated by set theory right from the beginning. He had also, around the turn of the century, published his *Grundlagen [Foundations] der Geometrie*, which used the example of geometry to argue for the use of formal axiomatics as a way of ensuring that a theory was a solid structure rather than a fragile house of straw. Hilbert had been the mathematician to whom Cantor first turned when he began to be faced with the several paradoxes that plagued him in that period. In fact, Zermelo had himself discovered Russell's paradox several years prior to Russell, who published it in 1903 and had informed Hilbert of it. Gregory H. Moore, the author of the excellent *Zermelo's Axiom of Choice*, points out, "But Zermelo did not publish it. This . . . suggests strongly that the paradox was much less compelling to him than it was to Russell, perhaps because Zermelo was more mathematically and less philosophically inclined."[9]

Hilbert therefore became interested in proving that the real numbers constituted a rigorous set; this would be a useful foundation for set theory, for it would show whether the set of real numbers could be properly placed among the alephs and would help show the consistency of the continuum hypothesis.* It remained, however, an unfinished work.

In the meantime, Zermelo, inspired by the objections and the comments from Borel and his group, had been moving in a similar direction. He too wanted to put set theory on a more solid foundation, while at the same time securing his demonstration of the well-ordering theorem and saving his axiom of choice. In addition, though of less importance, he wanted to deal with the current paradoxes, plus those that might arise later on. His approach would be similar to Hilbert's but further along—an axiomatization of set theory.

During the summer of 1907, he completed two important papers: a second, revised proof of the still very controversial well-ordering theorem and his axiomatization of set theory. These were both published in the same issue of *Mathematische Annalen* in 1908.

The revised proof of the well-ordering theorem used the axiom of choice and showed that the two are equivalent. Zermelo defended his use of the axiom and maintained that mathematicians should continue to use it unless and until it leads to contradictions. The axiom, he insisted, "has a purely objective character which is immediately clear."[10]

The first paper also contained a defense of his axiom of choice. He admitted that it had not been proved but argued that "in mathematics unprovability . . . is in no way equivalent to nonvalidity, since, after all, not everything can be proved, but every proof in turn presupposes unproved principles."[11] What, then, is the justification for use of the axiom of choice? He argued that it was necessary because it was being used to prove important theorems, and because it was necessary for science.

He argued that it had actually already found use in mathematics: "That this axiom, even though it was never formulated in textbook style, has frequently been used, and successfully at that, in the most diverse fields of mathematics . . . is an indisputable fact. . . . Such an extensive use of a principle can be explained only by its self-evidence."[12]

---

*A mathematical theory or system is consistent when none of its parts is inconsistent with, or contradicts, any other part within it.

In fact, as Penelope Maddy, the author of *Realism in Mathematics*, points out, "One of the great ironies of this entire historical episode is that the strongest negative reaction to the axiom came from the very group of French analysts–Baire, Borel, and Lebesgue–who unwittingly used it with great frequency and whose work provides part of the basic indispensability argument."[13]

Zermelo felt that his second paper, the axiomatization of set theory, was particularly important. There was, of course, his belief in the fundamental importance of set theory; he stated this right up front, calling it an indispensable component of all mathematics. As for the paradoxes that seemed to endanger set theory, his axiomatization, he believed, would go a long way toward providing an answer.

One problem had been that Cantor, in setting up what would later be referred to as his "naive" set theory, had not carefully restricted the concept of a set. Zermelo hoped that by using specific axioms, he could clarify the concept of a set. No properties of sets could be used unless specifically granted by the axioms. His plan was to include in the axioms only those sets and classes that seemed least likely to lead to paradoxes. Amazingly, he was able to set up his system to include just seven axioms, of which one was the axiom of choice! He had hoped to prove that his set was self-consistent before publication, but he was not able to do this. He decided to publish nonetheless.

Again Borel and his group criticized Zermelo's work, but the steam built up slowly. Initially, for example, Borel went along with some of Zermelo's and Hadamard's reasoning. He stated, "Of course it is possible to reason about a class of mathematical objects, for example all the real numbers or all the continuous functions; this class is defined by means of a finite number of words, although not *all* the members can be defined in that way. Thus one obtains the general properties of the class."[14]

Hadamard answered Borel, "I doubt that I have ever said otherwise. For indeed, all the members must exist in some way in order to form the class. . . . Zermelo would thus have demonstrated, if not that there exists one way of well-ordering the continuum, at least that there exists a [non-empty] *class* of such orderings. . . . This means, in sum, that (if Zermelo's argument is not ultimately perfected) we shall only be able to reason about the properties common to *all* these orderings. I willingly believe it. There are so many other things that we shall never know."[15]

Gregory Moore adds, "While granting Zermelo the right to endow any abstract entities with any non-contradictory properties that he wished, Borel stressed that such formal logic led to nothing but purely verbal conclusions, unrelated to reality. Despite Borel's scepticism, Hadamard's distinction between an object which can be defined uniquely and a non-empty class of objects, no one of which can be defined uniquely, would later assume considerable importance in mathematical logic."[16]

Hadamard continued to support Zermelo and to argue with others in support of the axiom of choice. Zermelo believed that his axioms were independent of each other; he also felt that the consistency of the system was a complex matter that remained to be established. Yet he felt that he had managed to provide an answer to the paradoxes.

The general reaction was that he had provided an improvement over Cantor's set theory, but that his system still needed work. Moore feels, "During the transitional decade 1909–1919, the disputed axiom of choice proved to be more secure than the axiomatic system that he introduced to serve as its foundation."[17] One of the arguments against Zermelo's system was that he had given no specific rationale for the axioms in it.

The arguments about the axiom and Zermelo's first proof of the well-ordering theorem did not seem to change much in this time either. Those who opposed the proof, including Borel, Lebesgue, and Russell, did so because of the nonconstructive character of the axiom that served as its foundation. Lebesgue and others saw in the axiom a process of reasoning with an infinite number of premises, which was not acceptable to them.

More Work

Bertrand Russell, whose paradox was so instrumental in driving much of this ongoing fixing-up work, never doubted the basic importance of set theory. He wrote in the preface to his monumental

Principia Mathematica (1910–1913, written with Alfred North White-head) that, aside from the symbolism, it was based entirely on the work of Georg Cantor.[18]

On the other hand, Jules Henri Poincaré, an important French mathematician, came to feel that the paradoxes, and especially Russell's paradox, showed clearly that set theory was a grave disease that could infect all mathematics. The two men, Russell and Poincaré, ended up in serious conflict over another matter, which we look at in the next chapter.

In the meantime, although Borel was critical of set theory, he was apparently more accepting of the basic idea than was Poincaré. Borel looked at set theory as somewhat analogous to mathematical physics. That is, it was not something real in itself, but it might be taken as a kind of guide that in turn could be used to discover new results, which would then have to be verified by accepted methods.[19]

The situation hardened into a standoff. Morris Kline writes:

The axiom [of choice] became a serious bone of contention.

Despite this, however, many mathematicians continued to use it as mathematics expanded in the succeeding decades. A conflict continued to rage among mathematicians about whether it was legitimate, acceptable mathematics. It became the most discussed axiom next to Euclid's parallel axiom. As Lebesgue remarked, the opponents could do no better than insult each other because there was no agreement. He himself, despite his negative and distrustful attitude toward the axiom, employed it, as he put it, audaciously and cautiously. He maintained that future developments would help us decide.[20]

This was a solid prediction, though he would surely be surprised at how things turned out. First, in 1921–1922, Abraham Adolph Fraenkel (1891–1965), seeing that Zermelo's axioms were insufficient tools for building all the sets needed for full use of set theory (for example, questions about the set of all sets), improved on Zermelo's work. He placed some restrictions on the formation of sets, needed to avoid paradoxes, but at the same time admitted enough sets for most classical analysis needs. Subsequently, a few modifications were made by others, but the resulting system of axioms became known as the Zermelo-Fraenkel system and came to be used widely by set theorists.

Incompleteness

While some mathematicians went about their business, others were seriously troubled by the paradoxes—the elephant in the garden. Yet Zermelo had hoped to do more than resolve the paradox problem. He, and others such as Hilbert, had felt that a solid axiomatization of set theory would provide a sound foundation for the theory of arithmetic and, in fact, for mathematics in general. Things have not turned out that way.

In the early 1930s, a young Austrian-born mathematician, Kurt Gödel, presented some work that showed, essentially, that such an axiomatization, no matter how carefully carried out, might never be able to provide the desired solid foundation. Gödel's *incompleteness theory* says that given any system, there will always be propositions that cannot be proven within the system. In other words, it is impossible to establish the consistency of set theory while working entirely within the system, no matter how it is established. Consistency can be accomplished only by the use of higher or external principles.

Joseph Dauben explains, "Gödel showed that in any system rich enough to contain elementary arithmetic, there were always theorems that could neither be proven nor disproven. They were undecidable, and it seemed quite possible that Cantor's continuum hypothesis might be a prominent example of such an undecidable proposition."[21]

Later, in 1963, Paul J. Cohen, a Stanford University mathematician, proved something that Gödel's work had suggested. Cohen showed that neither the continuum hypothesis nor the axiom of choice could be proved correct from within set theory, including the axiomatized arrangement of Zermelo-Fraenkel.

Cohen, in fact, believed further that the continuum hypothesis was actually false—that there *can* be a transfinite number between $\aleph_0$ and c—and that mathematicians would one day be able to show this. This, of course, would have broken Cantor's heart. Cohen's thinking went like this: he wondered why a rich concept like the continuum (c, and sometimes given as $2^{\aleph_0}$) should be equivalent in power to something as simple as the class $\aleph_1$. He felt further that the continuum might turn out to be larger than any transfinite aleph.

In any case, we had an answer (of sorts) to Hilbert's first problem (the proof of Cantor's continuum hypothesis). And clearly part of the answer depended on what axioms were chosen initially as the basis on which to work.

Cantor's set theory and the various, often quite serious, objections to it had brought mathematics to a difficult pass. Certainly, the long-held and cherished view of mathematics as a logical, exact, and certain discipline had been badly mauled. Not only did various mathematicians see things differently, they also began to group together and to take stands antipathetic to one another.

The paradoxes, for example, had a variety of effects, especially on people interested in the foundations of mathematics, for it began to appear that the whole structure of their beloved discipline was shaky or perhaps was built on a weak foundation. Starting around the turn of the 20th century, a fairly large number of mathematicians became engaged in studies along this line but divided into several mutually antagonistic groups. These formed gradually into three main groups, or schools: the logicists, headed by Bertrand Russell and Alfred North Whitehead; the intuitionists, founded by Leopold Kronecker, given some support by Jules Henri Poincaré, and then championed by Luitzen Brouwer and Hermann Weyl; and the formalists, under David Hilbert. We'll look at the latter two systems in a later chapter. In the next chapter, we consider Russell's logicism and how he came to it, along with Poincaré's quarrel with it, and how he came to that.

8

Poincaré versus Russell

The Logical Foundations
of Mathematics

When, in the spring of 1901, mathematicians were faced with Russell's paradox (which we discussed in chapter 6), many could feel the foundations of their discipline shaking under their feet. Subsequently Russell wrote:

> Philosophers and mathematicians reacted in various different ways to this situation. Poincaré, who disliked mathematical logic and had accused it of being sterile, exclaimed with glee, "It is no longer [merely] sterile, it begets contradiction." This was all very well, but it did nothing toward the solution of the problem. Some other mathematicians, who disapproved of Georg Cantor,

adopted the March Hare's solution: "I'm tired of this. Let's change the subject." This, also, appeared to me inadequate. After a time, however, there came to be serious attempts at solution by men who understood mathematical logic and realized the imperative necessity of a solution in terms of logic. The first of these was F. P. Ramsey, whose early death unfortunately left his work incomplete. But during the years before the publication of *Principia Mathematica* [three volumes, 1910–1913, by Russell and A. N. Whitehead], I did not have the advantage of these later attempts at solution, and was left virtually alone with my bewilderment.[1]

Following is Russell's own explanation of how he came upon the paradox. Remember, this is Russell talking, so don't worry too much if the logic escapes you on first, or second, reading. He was led to the contradiction, he wrote:

by considering Cantor's proof that there is no greatest cardinal number. I thought, in my innocence, that the number of all the things there are in the world must be the greatest possible number, and I applied his proof to this number to see what would happen. This process led me to the consideration of a very peculiar class.* Thinking along the lines which had hitherto seemed adequate, it seemed to me that a class sometimes is, and sometimes is not, a member of itself. The class of teaspoons, for example, is not another teaspoon, but the class of things that are not teaspoons. [In other words, the set of all teaspoons is not a teaspoon; so it is not a member of itself.] There seemed to be instances which are not negative: for example, the class of all classes is a class. The application of Cantor's argument led me to consider the classes that are not members of themselves; and these, it seemed, must form a class. I asked myself whether this class is a member of itself or not.[2]

---

*Russell's term *class* has the same meaning as the modern word *set*.

Thus was the famous Russell's paradox born. It hardly seems like enough to shake the foundations of anything. But there it was. Not only would it have significant consequences in the world of mathematics (see chapters 6 and 7), but it would lead to more than a decade of intellectual turmoil and extraordinary effort on Russell's part. Although there was some early support from a very limited number of colleagues, a fair amount of his effort went into dealing with the critiques of a variety of his peers.

Russell, as you may have gathered from my first paragraph, was in favor of mathematical logic. In fact, he is often considered the founder of a movement, called logicism, that still has adherents today but which brought forth the objections. As Russell put it, the logicist wants "to show that all pure mathematics follows from purely logical premises and uses only concepts definable in logical terms."[3]

Logicism is sometimes seen as a two-part effort. First, it claims that all of mathematics can be translated into logical terms. Thus the vocabulary and the symbols of mathematics constitute a valid subset of the vocabulary and the symbols of logic. Second, it claims that all mathematical proofs can be reset as logical proofs; thus the theorems of mathematics also make up a proper subset of the theorems of logic.

By way of emphasizing that pure mathematics is made up of logical steps, Russell stated, "Pure mathematics consists entirely of assertions to the effect that if such and such a proposition is true of anything [for example, if p, then q], then such and such another proposition is true of that thing. It is essential not to discuss whether the first proposition is really true, and not to mention what the any-thing is of which it is supposed to be true. . . . Thus mathematics may be defined as the subject in which we never know what we are talk-ing about, nor whether what we are saying is true."[4]

It's not hard to believe that there were criticisms. As Russell put it in a later writing: "This thesis was, at first, unpopular, because logic is traditionally associated with philosophy and Aristotle, so that mathematicians felt it to be none of their business, and those who considered themselves logicians resented being asked to master a new and rather difficult mathematical technique."[5] Among the most persistent and most insistent of his critics was the highly respected French mathematician Jules Henri Poincaré. This is not too surprising in that Poincaré had been, after Kronecker's death in 1891, the prime opponent of Cantor's transfinite mathematics, and Russell was building much of his logicism edifice on the foundation supplied by Cantor's set theory.

The series of arguments and counterarguments that passed between Russell and Poincaré ran from early 1906 through Russell's final reply in 1910. Russell was in his thirties and Poincaré was in his fifties during this period. By then, both men were highly esteemed by all in their fields, so each treated the other with respect. Charles Nordmann, a French astronomer, maintained in his eulogy of Poincaré that "among a dozen great scientists who lived during the last century, he accomplished the miracle of never having made a single enemy, a single one hostile to him in science."[6] Yet Poincaré and Russell pulled no punches in their intellectual criticisms of each other.

Before we get into the battle itself, let's talk a bit more about Russell and his mathematical logic.

Russell

Bertrand Arthur William Russell was born on May 18, 1872, in Trelleck, Wales. He lost his mother at age two and his father at age four. Brought up mainly by a grandmother (his grandfather died when he was six), he was educated at home by a succession of tutors until he was 18.

Although he loved and respected his grandmother for her good qualities, including her love for him and some progressive social inclinations, he began, by his adolescent years, to feel hemmed in. As he put it: "After I reached the age of fourteen, my grandmother's intellectual limitations became trying to me, and her Puritan morality began to seem excessive."[7] In fact, throughout his life Russell often found himself in conflicted intellectual and emotional situations.

By his teens, his intellectual powers were already in evidence. Russell wrote in his *Autobiography*,

At the age of eleven, I began Euclid, with my brother [seven years older than Bertrand] as my tutor. This was one of the great events of my life, as dazzling as first love. I had not imagined that there was anything so delicious in the world. After I had learned the fifth proposition,* my brother told me that it was generally considered difficult, but I had found no difficulty whatever. This was the first time it had dawned upon me that I might have some intelligence. From that moment until Whitehead and I finished *Principia Mathematica* when I was thirty-eight, mathematics was my chief interest, and my chief source of happiness. Like all happiness, however, it was not unalloyed. I had been told that Euclid proved things, and was much disappointed that he started with axioms. At first I refused to accept them unless my brother could offer me some reason for doing so, but he said: "If you don't accept them we cannot go on," and as I wished to go on, I reluctantly admitted them *pro tem*. The doubt as to the premisses of mathematics which I felt at that moment remained with me, and determined the course of my subsequent work.[8]

In 1890, he entered Trinity College, Cambridge, and studied mathematics and philosophy. Two years later, he was invited to join the Apostles. A small, highly select group that met at the university, it included A. N. Whitehead, then a lecturer in mathematics, who was to play a large part in Russell's future. Well aware of their intellectual prowess, the Apostles nevertheless managed not to take themselves too seriously. Russell saw this connection as the "greatest happiness of my life at Cambridge."[9] He felt, in fact, that he had gotten far more from the Apostles than from the establishment. The dons, he wrote, "contributed little to my enjoyment of Cambridge,"[10] and he added that he "derived no benefits from lectures."[11]

Here's more on his early development:

I had already been interested in philosophy before I went to Cambridge, but I had not read much except Mill. What I most desired was to find some reason for supposing mathematics true. The

---

*In isosceles triangles, the angles at the base are equal to one another, and, if the equal straight lines are produced further, the angles under the base will be equal to one another.

arguments in Mill's *Logic* on this subject already struck me as very inadequate. . . . My mathematical tutors had never shown me any reason to suppose the Calculus anything but a tissue of fallacies. . . . During my fourth year I read most of the great philosophers as well as masses of books on the philosophy of mathematics. James Ward [Russell's tutor at Cambridge] was always giving me fresh books on this subject, and each time I returned them, saying they were very bad books. I remember his disappointment, and his painstaking endeavors to find some book that would satisfy me. In the end, but after I had become a Fellow, I got from him two small books, neither of which he had read or supposed of any value. They were Georg Cantor's *Mannichfaltigkeitslehre*,[12] and Frege's *Begriffsschrift*.[13] These two books at last gave me the gist of what I wanted."[14] [We'll discuss Frege later in this chapter. For the moment, let's just say that the *Begriffsschrift*, a work on logic, included a formal language on which to found arithmetic.]

Russell's fascination with Cantor had some curious byways. In the last years of the 19th century, Russell was lecturing at the London School of Economics. He later wrote, "I used to walk every day to [my wife's] parents' house in Grosvenor Road, where I spent the time reading Georg Cantor, and copying out the gist of him into a notebook. At that time I falsely supposed all his arguments to be fallacious, but I nevertheless went through them all in the minutest detail. This stood me in good stead when later on I discovered that all the fallacies were mine."[15]

At the time of Russell's attendance, the university was undergoing a significant change, in that the administration was beginning to see intellectual/academic research as an important part of the teachers' work, rather than as something to be done after classroom hours. Original research work could lead to multiyear fellowships, and indeed Russell earned one in 1895 for a dissertation on the foundations of geometry. It was published in 1897.

Following this success, Russell began putting together ideas for a comprehensive treatment of the principles of mathematics, but one with a particular slant. His studies and, mainly, his discussions with his friends and lecturers had led him to begin thinking that it would be possible to create a mathematics based on a small number of fundamental logical concepts. Here were the beginnings of Russell's logicism.

Logicism

It's important to understand, as Russell surely did, that mathematical logic and even logicism did not spring *de novo* from the mind and the pen of Bertrand Russell. In chapter 3 of this book, for example, I mentioned Leibniz's interest in using symbolic logic to create a kind of calculus of reasoning. In subsequent years, many mathematicians looked into and worked with various aspects of logic including mathematical logic and the foundations of mathematics, but two had a particular influence on the direction of Russell's research.

By the late 1870s, Gottlob Frege, a German logician/mathematician/ philosopher, had found that much of mathematics could be derived from a much smaller set of logical statements. He had already, in 1884, published the *Grundlagen [Foundations] der Arithmetik*, which was an early attempt at an axiomatization of arithmetic. The book was mostly ignored; the only recorded review is one by our own Georg Cantor who, apparently, did not really understand it and gave it a blistering review.

Logicism—the intimate mating of logic and mathematics—was theoretically possible, Frege believed, and he began working out the derivations that would be needed. By 1902, he had pulled his work together and had already published the first of what was to be two volumes of his *Grundgesetze [Basic Laws] der Arithmetik*. He was just at the point of putting his second volume to press when Russell, who had been impressed by Frege's earlier *Begriffsschrift* (1879), realized that his own paradox created a contradiction in Frege's system of axioms. Russell pointed this out to Frege in a letter (June 16, 1902). Frege was devastated.

It was too late for Frege to make any changes in his text, for the pages had already been printed, but he did add an appendix, which began with the startling statement: "A scientist can hardly meet with anything more undesirable than to have the foundation give way just as the work is finished. I was put in this position by a letter from Mr. Bertrand Russell when the work was nearly through the press."[16] Frege modified the axiom in the appendix, though this created problems elsewhere in the text and especially in the first volume.

History records that Frege became seriously depressed after this, though the reasons were actually mainly personal and even political. In later years, he recovered and began to do some fine work again, though never again in this field. By 1923, he actually came to the conclusion that the attempt to found mathematics on logic was misguided.

The irony here is that Russell had begun work on the first major explication of his own efforts in logicism, *The Principles of Mathematics* (1903), when he came up with his paradox in 1901. Although Frege simply gave up in his attempt to derive mathematics from logic, Russell did not; he decided to go on and publish his work—including the paradox, without having an answer to it—and to continue to seek a solution. Frege's second volume was published, too, but not until 10 years later; a third projected volume never was done.

Yet in the preface to *Principles*, Russell himself admits, "Professor Frege's work, which largely anticipates my own, was for the most part unknown to me when the printing of the present work began; I had seen his *Grundgesetze der Arithmetik*, but, owing to the great difficulty of his symbolism, I had failed to grasp its importance or to understand its contents. The only method, at so late a stage, of doing justice to his work, was to devote an Appendix to it."[17] In other words, Frege was on the right track, but he felt that Russell's paradox derailed him. This left Russell an open track, though a far from easy one.

Another irony is that, as Russell himself later stated, "In spite of the epoch-making nature of [Frege's] discoveries, he remained wholly without recognition until I drew attention to him in 1903."[18]

(Were Frege to come to life today, he would be proud, and probably astounded, to learn that neo-Fregeanism is the order of the day. There have in recent decades been serious explorations of his work and attempts to incorporate it into current efforts and applications.[19])

Russell's objective was to create a more comprehensive treatment of the principles of mathematics. He had come to believe more strongly that pure mathematics could be built on a small group of fundamental logical concepts and that all its propositions could be deduced from a small number of basic logical principles, but he was not pleased with his early drafts.

In 1900, however, he had attended the International Congress of Philosophy in Paris, a meeting that, as he later put it, "was a turning point in my intellectual life, because there I met [Giuseppi] Peano. . . . In discussions at the Congress I observed that he was always more precise than anyone else, and that he invariably got the better of any argument upon which he embarked. As the days went by, I decided that this must be owing to his mathematical logic. I therefore got him to give me all his works, and as soon as the Congress was over I retired to Fernhurst [Russell's home base] to study quietly every word written by him and his disciples. It became clear to me that his notation afforded an instrument of logical analysis such as I had been seeking for years."[20] Examples are the use of $\supset$ for "implies" or "contains," and $\in$ for "belongs to." Thus the phrase "The entity y is a member of the class A" is replaced by $y \in A$. The result is both brief and precise. With this symbolism, Peano had managed to express definitions, theorems, and proofs.

Russell quickly ingested Peano's ideas and symbolism, which was quite comprehensive, and began rewriting his book, which would build on Peano's work. For example, as Russell explained later, "Having reduced all traditional pure mathematics to the theory of natural numbers, the next step in logical analysis was to reduce this theory itself to the smallest set of premises and undefined terms from which it could be derived. This work was accomplished by Peano. He showed that the entire theory of the natural numbers could be derived from three primitive ideas and five primitive propositions in addition to those of pure logic. These three ideas and five propositions thus became, as it were, hostages for the whole of traditional pure mathematics. If they could be defined and proved in terms of others, so could all of pure mathematics."[21]

Logicism Emerges, Warts and All

Russell's *Principles of Mathematics* was to appear in two volumes. Although Russell had not been happy with the first drafts of *Principles*, the writing went smoothly in the months after the Paris meeting. Volume 1, which appeared in 1903, was essentially a popular work, presenting the basic ideas supporting the connections between logic and mathematics. Volume 2 was to contain the necessary proofs, but it was never done. Or, rather, it morphed into the massive, three-volume *Principia Mathematica*, which he did in stages, with the collaboration of his friend and colleague Alfred North Whitehead (1861–1947).

This massive work, consisting of more than 2,000 pages, was and still is regarded as one of the major works of mathematical literature. Though not often read, it stood out as Russell's brief for the idea that mathematics derives from the rules of logic, and it gave the premises that showed how these rules were used in number theory, in set theory, and in other areas of mathematics.

Russell later threw out a challenge: "If there are still those who do not admit the identity of logic and mathematics, we may challenge them to indicate at what point in the successive definitions and deductions of *Principia Mathematica* they consider that logic ends and mathematics begins."[22]

It was a Herculean job. The authors had estimated the job might take a year to complete, but the first volume did not appear until 1910, the third and final volume in 1913. At one point, starting in 1907, Russell was working 10 hours a day, 8 months a year, for the next 3 years. At the publication of volume 1, another problem arose. Cambridge University Press, which had contracted to publish the work, was not expecting the mass of pages that would be coming its way. Publication, it seemed, would lead to a loss of 600 pounds. The firm agreed to absorb half this amount, if the authors could raise the other half. The Royal Society of London, to its credit, donated 200 pounds, and the authors each contributed 50 pounds.

In the meantime, however, Russell was still trying to deal with the

paradoxes. These contradictions, he was coming to suspect, constituted a kind of vicious circle, and he sought ways to get around this.
He had tried an early attempt at a method called the theory of types.
The basic idea was to distinguish between individuals, ranges of
individuals, ranges of ranges of individuals, and so on. Each level
was called a type, and he stipulated that if the expression *x is a u* is
to have meaning, then *u* must be one level or type higher than *x*. He
had included this in the appendix to his *Principles of Mathematics*.
Though the concept had been kicking around for several years, this
was the first time it had appeared in print; but while it could deal
with his paradox, it couldn't handle Cantor's, so he was not really
happy with it.

Ironically, Poincaré had also tried to handle the paradoxes with a
similar idea. He felt that the paradoxes involved a collection and a
member of it whose definition depended on the collection as a unit.
This kind of definition, which he called impredicative, was similar in
concept to what Russell called a vicious circle. Eliminating such sets
would prevent troublesome paradoxes. This worked but would place
a severe restriction on the procedure. The main problem with it was
that many established aspects of mathematical activity were founded
on just this kind of collection.

In 1905, Russell tried again with some new ideas. At this point,
he developed three different approaches to the problem: the zig-zag
approach set limits on how complex a propositional function can be
when considered as a defining class; the limitation-of-size theory set
up rules that prevent certain classes from becoming too large and
thereby creating contradictions; and the no-classes theory proposed
to do away with classes altogether. Each of these methods became the
subject of future investigations. He presented them in a paper, "On
Some Difficulties in the Theory of Transfinite Types and Order
Types," which he read to the London Mathematical Society on
December 14, 1905. It was published in the Society's *Proceedings*
on March 7, 1906. He began with the comment: "Each of the three
theories can be recommended as plausible by the help of certain *a
priori* logical considerations."[23]

Poincaré

Watching these goings-on from the sidelines was the respected French mathematician Jules Henri Poincaré. Born in Nancy, France, on April 29, 1854, he was 18 years older than Russell. One of the few broad-based mathematical scientists in an era of growing specialization, he had by the turn of the century already built a reputation in several fields, including number theory, topology, probability, and various areas of mathematical physics, plus having written a still-famous three-volume work on celestial mechanics. He even did pioneering work in the special theory of relativity and later made important contributions in the philosophy of science.

He passed his young years amid a circle of intellectuals and high achievers. After a short period of education at home, he attended the Nancy lycée, the École Polytechnique, the School of Mines, and then the University of Paris, where he received his doctorate in mathematical sciences in 1879. His thesis was on differential equations. In 1881, he became a lecturer and in 1886 was made a full professor at the University of Paris, where he remained until his death in 1912.

All the way through his young years, his brilliance set him apart from his neighbors and peers, though his weak eyesight, physical frailty, and poor coordination opened him to the teasing and bullying that one might expect in these circumstances.

His mathematical and scientific output, however, starting immediately after his schooling ended, was impressive: he published more than 30 books on mathematical physics and nearly 500 papers on mathematics. He also wrote popular essays and three volumes on the philosophy of science that are considered classics in their field. James R. Newman, a well-known historian of science, makes an interesting point about Poincaré's writing style: "Except for the Gallic flavor of his sentences, his liquid, subtle style resembles Bertrand Russell's."[24] We'll be able to compare samples of their writing shortly.

Certain specifics of Poincaré's working methods give some indication of the man. He kept very specific working hours—from 10 A.M. to noon and from 5 to 7 P.M. He read his journals in the evening.

Although he read widely, he tended not to build on the work of others. In his own work, he developed ideas straight through from the basic beginnings. This mode of operation began in his early years and explains why he was able to take all his mathematical courses at the École Polytechnique without writing a single note. It was not that he remembered everything; it was that he could reason things out as he needed them. E. Toulouse, one of his biographers, later maintained that often Poincaré did not make an overall plan when writing a paper—indeed, he might start without knowing where he would end.

By the time of the tiff with Russell, he had been honored with just about every medal and prize available and had been elected to membership in the most distinguished scientific and mathematical organizations. One of these honors seemed to create a turning point in his career. He was elected to membership in the Academy of Sciences in 1887 at the tender age of 32. This apparently led him toward a greater interest in interacting with the public, and he began writing for a wider audience. Of his close to 100 nontechnical books and articles, almost all were written after his election to the academy.

This coincided with the fact that his reputation, both at home and abroad, was growing, and he was being called upon to speak or write on topics of mathematics and science appropriate for a more general audience. Yet this was easy for him. As a mathematician and a scientist, he was unusual, in that he had an extraordinarily wide range of interests and was capable in all of them. He read widely and was aware of what was going on all around him. He also began to pay more attention to basic questions about the nature and the philosophy of mathematics.

Like Kronecker and others of his time, he had some very definite ideas on the new mathematical ideas taking root in his era. He felt, for example, that it was not necessary to define the whole numbers or axiomatize their properties; that one should not introduce objects that one cannot define clearly and completely in a finite number of words.

He called set theory a pathological case and predicted, "Later generations will regard [Cantor's] set theory as a disease from which one has recovered."[25]

He felt that some mathematical ideas were more basic than logic and could not be stated in logical terms. In 1904, he wrote, "It is by logic that we prove, it is by intuition that we invent." Later he stated, "Logic, therefore, remains barren unless fertilized by intuition."[26]

Of the kind of mathematics he believed in, it's no surprise that he leaned toward and worked mainly in applied mathematics. "Experience," he said, "is the sole source of all truth."[27] Though this eventually led him to think deeply about the foundations of scientific knowledge, his leanings toward the concrete held firm. Thus, in contrast with Cantor, who saw infinity as a real and workable concept, Poincaré argued against infinite sets. In fact, he maintained, "Actual infinity does not exist. What we call infinite is only the endless possibility of creating new objects no matter how many objects exist already."[28] He was also, writes Morris Kline, "totally antipathetic to the heavily symbolic logistic approach and in his *Science and Method* was even sarcastic. Referring to one such approach to the whole number advanced by [Cesare] Burali-Forti in an article of 1897 wherein one finds a maze of symbols that define the number 1, Poincaré remarked that this is a definition admirably suited to give an idea of the number 1 to people who never heard of it before."[29]

In another of Poincaré's earlier articles, he made one of his more extreme statements. "Logic," he wrote, "sometimes makes monsters. [For] half a century we have seen arise a crowd of bizarre functions which seem to try to resemble as little as possible the honest functions which serve some purpose. . . . Heretofore when a new function was invented, it was for some practical end; to-day they are invented expressly to put at fault the reasonings of our fathers, and one will never get from them anything more than that."[30]

So we are not surprised to find that Poincaré turned out to be the main opponent of logicism, which depends heavily on set theory. For a while, the chief French proponent of Russell's logicism was the French mathematician Louis Couturat, who published some articles in 1904 and 1906. Poincaré chose not to air his feelings. With the publication of Russell's 1906 article in the *Proceedings of the London Mathematical Society*, he decided it was time to act.

Response

It was at this point that Poincaré decided to mount a general attack on Russell's logicism. The French journal *Revue de Métaphysique et de Morale* had begun publication in 1893. Its objective was to bring philosophy and the various sciences (moral as well as natural) into mutual understanding, and Poincaré had become one of its main contributors. Thus he chose to launch his attack here. It appeared as "Les mathématiques et la logique" in May 1906, two short months after Russell's paper, which he used as the springboard.

He began with an attack that led back to Cantor. After giving a short introduction to Cantor's set theory, Poincaré wrote, "Many mathematicians followed his [Cantor's] lead. . . . In their eyes, to teach arithmetic in a way truly logical, one should begin by establishing the general properties of transfinite cardinal numbers, then distinguish among them a very small class, that of the ordinary whole numbers. Thanks to this detour, one might succeed in proving all the propositions relative to this little class (that is to say all our arithmetic and all our algebra) without using any principle foreign to logic." Poincaré maintained, however, that "This method is evidently contrary to all sane psychology; it is certainly not in this way that the human mind proceeded in constructing mathematics; so its authors do not dream, I think, of introducing it into secondary teaching. But is it at least logic, or, better, is it correct? It may be doubted. . . .

He went on,

> Unfortunately, they have reached contradictory results, what are called the *cantorian antinomies* [that is, the paradoxes]. . . . These contradictions have not discouraged them and they have tried to modify their rules so as to make those disappear which had already shown themselves, without being sure, for all that, that new ones would not manifest themselves.
>
> It is time to administer justice on these exaggerations. I do not hope to convince them; for they have lived too long in this atmosphere. Besides, when one of their demonstrations has been refuted, we are sure to see it resurrected with insignificant alterations, and some of them have already risen several times from their ashes.

Then:

Thus, be it understood, to demonstrate a theorem, it is neither necessary nor even advantageous to know what it means. The geometer might be replaced by the *logic piano* . . . ; or, if you choose, a machine might be imagined where the assumptions were put in at one end, while the theorems came out at the other, like the legendary Chicago machine where the pigs go in alive and come out transformed into hams and sausages. No more than these machines need the mathematician know what he does.

And then the logical correctness of the reasonings leading from the assumptions to the theorems is not the only thing which should occupy us. The rules of perfect logic, are they the whole of mathematics? As well say the whole art of playing chess reduces to the rules of the moves of the pieces. Among all the constructs which can be built up of the materials furnished by logic, choice must be made; the true geometer makes this choice judiciously because he is guided by a sure instinct, or by some vague consciousness of I know not what more profound and more hidden geometry; which alone gives value to the edifice constructed.[31]

Here are a few of Poincaré's comments about Russell's attempts to deal with the "contradictions": "According to the zigzag theory, 'definitions (propositional functions) determine a class when they are very simple and cease to do so when they are complicated and obscure.' Who, now, is to decide whether a definition may be regarded as simple enough to be acceptable? To this question there is no answer, if it be not the loyal avowal of a complete inability: [ironically quoting Russell:] 'The rules which enable us to recognize whether these definitions are predicative would be extremely complicated and can not commend themselves by any plausible reason.' . . . I have not been able to find any other directing principle than the absence of contradiction."

Poincaré ends this point with, "This theory therefore remains very obscure; in this night a single light—the word *zigzag*. What Russell calls the 'zigzaginess' is doubtless the particular characteristic which distinguishes the argument of Epimenides." (Poincaré is referring to the Epimenides statement "I am lying," which leads to a paradox. If he is lying, he is telling the truth; if he is telling the truth, he is lying.)

Of the limitation-of-size theory, Poincaré argues, "A class would cease to have the right to exist if it were too extended. Perhaps it might be infinite, but it should not be too much so. But we always meet again the same difficulty; at what precise moment does it begin to be too much so? Of course this difficulty is not solved and Russell passes on the third theory."[32]

Poincaré then moves on to Russell's no-classes theory. Note first, however, that at the end of Russell's *Proceedings* paper, Russell had tacked on this addendum: "From further investigation I now feel hardly any doubt that the no-classes theory affords the complete solution of all the difficulties stated in the first section of this paper."[33]

Poincaré does not quite agree. He charges, "In the no-classes theory it is forbidden to speak the word 'class' and this word must be replaced by various periphrases. What a change for logistic which talks only of classes and classes of classes! It becomes necessary to remake the whole of logistic. Imagine how a page of logistic would look upon suppressing all the propositions where it is a question of class. There would only be some scattered survivors in the midst of a blank page. *Apparent rari nantes in gurgite vasto.*" (Only here and there men are seen swimming in the immense whirlpool.)[34]

Before moving on to Russell's responses and counterattack, I'd like to add a few more of Poincaré's charges:

"On the question of fertility, it seems M. Couturat has naive illusions. Logistic, according to him, lends invention 'stilts and wings.'" And, on the next page: "Ten years ago, Peano published the first edition of his *Formulaire*. How is it that, ten years of wings and not to have flown!

"I have the highest regard for Peano, who has done very pretty things (for instance his 'space-filling curve,' a phrase now discarded); but after all he has not gone further nor higher nor quicker than the majority of wingless mathematicians, and would have done just as well with his legs.

"On the contrary I see in logistic only shackles for the inventor. It is no aid in conciseness—far from it, and if twenty-seven equations were necessary to establish that 1 is a number, how many would be needed to prove a real theorem?"[35]

Counterattack

To be sure that Poincaré got his point, Russell responded in Poincaré's home territory, the *Revue de Métaphysique et de Morale*. In the September 1906 issue, he began, "M. Poincaré's article in this review, 'Les mathématiques et la logique,' (May 1906) illustrates what I believe to be a misapprehension as to the nature and purposes of logistic. . . . At the same time, it suggests a solution of the paradoxes besetting the theory of the transfinite. M. Poincaré holds that these paradoxes all spring from some kind of vicious circle, and in this I agree with him. But he fails to realize the difficulty of avoiding a vicious circle of this sort. I shall try to show that, if it is to be avoided, something like my 'no-classes theory' seems necessary; indeed it was for this purpose that I invented the theory."[36] There follows some twenty pages of explanation, which also includes other answers to Poincaré's charges.

A particularly interesting example is his answer to Poincaré's disparagement of Peano. Russell answers, "A point in which I must venture to differ respectfully from M. Poincaré is his estimate of M. Peano. [Here Russell repeats Poincaré's charge that Peano 'has gone no further . . . than . . . wingless mathematicians, and he could have done all that by walking on the ground.']

"Now I would suggest to M. Poincaré that this is merely a way of stating that the bulk of what M. Peano has done does not interest him. M. Peano has forged an instrument of great potency for certain kinds of investigations. Some of us are interested in such investigations, and therefore do honour to M. Peano, who, as regards them, has gone, we think, so much farther and faster than the 'wingless' mathematicians that they have lost sight of him."[37]

Of Poincaré's comments about Russell's no-classes theory, Russell had this to say: "If M. Poincaré could divest himself of the belief that logistic is quite unlike any other part of mathematics, he would also realize that, in proposing not to regard classes as independent entities, I am not proposing a change which will make it necessary to 'remake all of logistic'; nor do I wish to forbid people to 'pronounce the word *class*' any more than Copernicus wished to forbid people speaking of the sunrise."

In other words, Poincaré's problem is that he simply does not understand what Russell is doing. "Perhaps," Russell wrote, "an analogy will make it clear that the change is not so great after all. The infinitesimal calculus, as is now universally recognized, neither employs nor assumes infinitesimals. But how much has this altered 'the appearance of one page of' the infinitesimal calculus? Hardly at all. Certain proofs are re-written, certain paradoxes which troubled the eighteenth century have been solved; otherwise, the formulae of the calculus have scarcely changed."[38]

Russell concludes, "M. Poincaré informs us that 'clearer notions in logic' are not what is wanted; but he does not reveal the process by which he has made this important discovery. For my part, I cannot but think that his attempts to avoid the vicious circle illustrate the fate of those who despise logic."[39]

And So It Goes

In another counterattack, Poincaré wrote, "*There is no actual . . . infinity.* The Cantorians have forgotten this, and they have fallen into contradiction. It is true that Cantorism has been of service, but this was when applied to a real problem whose terms were precisely defined. . . .

"Logistic also forgot it, like the Cantorians, and encountered the same difficulties." And later, "Russell has perceived the peril and takes counsel. He is about to change everything, and, what is easily understood, he is preparing not only to introduce new principles which shall allow of operations formerly forbidden, but he is preparing to forbid operations he formerly thought legitimate. Not content to adore what he has burned, he is about to burn what he adored, which is more serious. He does not add a new wing to the building, he saps its foundation."[40]

Russell answered in a new paper, titled "Mathematical Logic As Based on the Theory of Types," published in 1908 in the *American Journal of Mathematics*.[41] In it, he presented a new theory of types. Poincaré responded in 1909 with "La logique de l'infini" in the *Revue*. Here he gave what turned out to be his final suggestions for dealing with the paradoxes troubling logistics.[42]

1. Consider only objects that can be defined in a finite number of words;

2. Never forget that every proposition concerning infinity must be the translation, the abridged statement, of a proposition concerning the finite;

3. Avoid definitions and classifications that are not predicative.[43]

Regarding the distinction between the finite and the infinite, Poincaré stated in his 1909 article: "M. Russell will doubtless tell me that these are not matters of psychology, but of logic and epistemology. I shall be driven to respond that there is not logic and epistemology independent of psychology. This profession of faith will probably close the discussion, since it will show an irredeemable divergence of views."[44]

The issue was not closed, however, as far as Russell was concerned. He responded in May 1910 with his "La théorie des types logiques," again in Poincaré's home base, the *Revue*. By this time, the first volume of the *Principia Mathematica* was coming onto the scene. In its introduction, as in this latest article, he presented his latest thinking in logistic theory.

In the article he dealt, once again, with several subjects, including his agreement that "the paradoxes to be avoided all result from a certain kind of vicious circle." He added some new definitional work on classes.[45]

He then expanded on his earlier treatments. Later in the article, he once again explained his theory of classes.[46] Still later, he wrote, "One point in M. Poincaré's article on "La logique de l'infini" calls for a word of explanation. He [Poincaré] asserts (p. 469): 'The theory of types remains incomprehensible, unless one supposes the theory of ordinals already established.' This assertion appears to me [Russell] to rest upon a confusion"[47]—which Russell proceeded to try to clear up.

Would the interchange have continued? Perhaps—but fate intervened. Poincaré became ill with prostate problems not long afterward. He underwent an operation at a nursing home and seemed to make a good recovery, but complications set in and he died on July 12, 1912.

Russell, after Poincaré

What effect did the objections of Poincaré (and others) have on Russell and his ideas on logicism? We can find a fairly clear picture in the 1938 reissue of his 1903 *Principles of Mathematics*. Happily, he decided that "Such interest as the book now possesses is historical, and consists in the fact that it represents a certain stage in the development of its subject. I have therefore altered nothing, but shall endeavor, in this Introduction, to say in what respects I adhere to the opinions which it expresses, and in what other respects subsequent research seems to me to have shown them to be erroneous."

In sum, he tells us, "The fundamental thesis of the following pages, that mathematics and logic are identical, is one which I have never since seen any reason to modify." (That is, from 1903 to 1938.)[48] Some things, it seems however, continue to puzzle, including the very definition of logic: "To define logic, or mathematics, is therefore by no means easy except in relation to some given set of premisses."[49]

He also mentions Poincaré. Even in 1938, which was 26 years after Poincaré's death, Russell still felt it necessary to put a poultice on the sting of Poincaré's famous comment. Russell wrote, "I come finally to the question of the contradictions and the doctrine of types. Henri Poincaré, who considered mathematical logic to be no help in discovery, and therefore sterile, rejoiced in the contradictions: 'La logistique n'est plus stérile; elle engendre las contradiction!' All that mathematical logic did, however, was to make it evident that contradictions follow from premises previously accepted by all logicians, however innocent of mathematics. Nor were the contradictions all new; some dated from Greek times."[50]

Yet Russell was not so foolish as to think, or suggest, that nothing had changed in those years. He admits, later in the introduction, "There are still many controversial questions in mathematical logic, which . . . I have made no attempt to solve. I have mentioned only those matters as to which, in my opinion, there has been some fairly definite advance since the time when the 'Principles' was written [1900–1903]. . . . The changes in philosophy which seem to me to be called for are partly due to the technical advances of mathematical logic in the intervening thirty-four years."[51]

Certainly, there were changes. As Kline points out, "Although Russell and Whitehead had no hesitation in introducing the axioms of infinity and choice in the first edition of their *Principia Mathematica*, they certainly backtracked later, not only in acknowledging that the primary laws of logic were not absolute truths but also in recognizing that these two axioms are not axioms of logic. In the second edition of the *Principia*, these axioms were not listed at the outset and their use where needed to prove certain theorems was specifically mentioned."[52]

In fact, in these later years, Russell was no longer as confident in the ultimate success of his ideas as he was in his more optimistic early years. He doesn't say so in the 1938 Introduction to his *Principles*, but part of the reason could surely be Godel's 1931 proof of the incompatibility of consistency and completeness. (See chapter 7.) This could have led to a weakening of the promise of logicism as it was known in its early years.

As Russell put it, however, the reasons for doubt were, "broadly speaking, of two opposite kinds: first, that there are certain unsolved difficulties in mathematical logic, which make it appear less certain than mathematics is believed to be; and secondly that, if the logical basis of mathematics is accepted, it justifies, or tends to justify, much work, such as that of Georg Cantor, which is viewed with suspicion by many mathematicians on account of the unsolved paradoxes which it shares with logic. These two opposite lines of criticism are represented by the formalists, led by Hilbert, and the intuitionists, led by [Luitzen] Brouwer."[53]

In the next chapter, we consider these two schools of mathematical thought, their connections with logicism, and the part they played in the crisis of confidence that beset mathematics in the early years of the 20th century.

Today there is possibly as much controversy surrounding Russell's logicism as there ever was. Michael Detlefsen, at the University of Notre Dame, argues, for example, that "Russell's alleged refutation of Poincaré's Kantian viewpoint is mistaken." He maintains, "In the end, we find both the logicist's claim that mathematical reasoning can be 'logicized' and the claim that this is required for the perfection of rigor to be ill founded."[54] Some researchers feel that logicism remains too confused or too feeble to be of much use.[55] Others believe that with appropriate fixes, it remains a feasible technique.[56]

There can be little argument, however, that in one form or another, Russell's logicism led to research and advances, from his day to ours, in such varied areas as philosophy, mathematics, linguistics, and economics and, more and more today, in computers.[57]

9

Hilbert versus Brouwer

Formalism versus Intuitionism

In the last chapter, you read about contradictions, paradoxes, and a crisis that shook mathematics to its very foundations. The word *crisis*, however, should not be interpreted too broadly. The *methodology* of mathematics was never in question; its practitioners and users continued throughout these years—that is, the early years of the 20th century—to put their mathematical techniques to very good use. This was especially true in the sciences—which saw remarkable advances in relativity and quantum theories—and in the humanities and other areas as well.

Even the difficulties caused by the paradoxes found in set theory were largely overcome, once it was determined that all that was needed was the application of some not too onerous restrictions.

In other words, the difficulties and the more serious arguments were not with mathematical technique. Rather, they had to do with the foundations of mathematics and with questions about the limitations of mathematical knowledge. These were given a certain urgency by some of the developments we have discussed in earlier chapters.

For a while, it seemed that Bertrand Russell's logicism (see chapter 8) might provide the answers to some or all of the questions bedeviling the questioners—indeed, that it might even provide the necessary basis for a solid foundation of mathematics. With the publication of the third great volume of *Principia Mathematica* in 1913, however, the movement began to falter. The books were widely admired but not often read. Russell's logicism had been able to get around the various paradoxes with his various devices, but he had not been able to show that his system would remain free of contradictions.

A planned fourth volume—to deal specifically with the logical foundations of geometry and to be written mainly by A. N. Whitehead—was interrupted by the advent of World War I. Russell turned his attention to pacifist activities and got himself into trouble with the British government. In 1918, he was sentenced to six months in prison for an article libeling the American army. There he wrote his *Introduction to Mathematical Philosophy* (1919), which was a kind of introduction to the *Principia*, that is, an attempt to make his logicistic ideas more accessible not only to specialists but also to "that wider circle who feel a desire to know the bearings of this important modern science."[1] Whitehead, more inclined to philosophy anyway, went off to teach in the United States in 1924.

In the meantime, the mathematical world had had time to digest David Hilbert's masterful *Grundlagen der Geometrie* (*Foundations of Geometry*). Published in 1898–1899, the work was later translated into the major European languages and was to prove extremely influential.

By the early years of the new century, Hilbert had built a strong reputation, having already done pioneering work in invariant theory and the calculus of variations, as well as having produced an important and influential book on the foundations of geometry. He was elected to membership in foreign academies and was awarded the title of *Geheimrat* by the German government, the equivalent of a knighthood in England.

Beginnings of Formalism

Invited to speak at the Second International Congress of Mathematicians (Paris 1900), Hilbert presented his list of the most challenging problems facing mathematicians of that day. Of the 10 problems he talked about at the lecture, the first 3 concerned the foundations of mathematics. The first, which I discussed in chapter 7, called for a proof of Cantor's continuum hypothesis. The second sought a proof of the consistency of the axioms of arithmetic—that is, proof that a finite number of logical steps based upon these axioms can never lead to contradictory results. By extension, he was questioning the very foundations of mathematics itself. As he put it in this same lecture, "The proof of the compatibility of the axioms [of arithmetic] is at the same time the proof of the mathematical existence of the complete system of real numbers or of the continuum."[2]

The third problem was to axiomatize those physical sciences in which mathematics plays an important role.[3]

By the time of the Third Congress, four years later, the emergence of the various paradoxes (chapters 7 and 8) had thrown a feeling of uncertainty regarding these foundational questions into the minds of many mathematicians.

As Hilbert saw it: "The present state of affairs is intolerable. Just think, the definitions and deductive methods which everyone learns, teaches and uses in mathematics, the paragon of certitude and truth, lead to absurdities! If mathematical thinking is defective, where are we to find truth and certitude?"[4] Hoping to build on the success he had achieved with his axiomatization of geometry, he thought, Why not apply the same approach to all of mathematics? He suggested at the conference, "I believe that all the difficulties that I have touched upon may be overcome, and an entirely satisfactory foundation of the number concept can be reached by a method which I call the axiomatic method, and whose leading idea I wish now to develop."[5]

Among his ideas was a desire to make axiomatic systems more general. He wanted to establish the self-consistency of the axioms of arithmetic and the steps that derive from them. He also felt that the paradoxes that posed such a problem to Russell, et al., were due to the semantic content of the language used, that is, to the vagueness of words. Thus, for example, Hilbert maintained, starting as early as 1891, "It must be possible to replace in all geometric statements the words *point, line, plane* by *table, chair, beer mug*."[6] His objective was to be able to deal with the mathematical symbols according to the specified or formal rules without need for a "meaning" (physical, abstract, or intuitive) for the symbols. A suggestive, if not wholly accurate, analogy is that of chess pieces; their names are suggestive but do not determine the action of chess. The action depends completely on the conventions and the rules of the game.

Thus was born the beginnings of Hilbert's school called formalism. The term *formalism*, interestingly, was first used by Brouwer. In fact, he used it specifically for Hilbert-type formalism. He lumped Cantor and Frege-Russell (see chapter 8) under the name *classical* (that is, non-intuitionistic) mathematics.

Furthermore, others had also done some earlier work in what became known as the formalist position, including Poincaré and Couturat.[7] It was Hilbert, however, who pulled the pieces together to the point where it could be called a school with a name and followers. By now, after all, he was one of the foremost mathematicians in Europe—his fame was exceeded only by that of Poincaré—so that anything he maintained was of interest.

In the early years of the 20th century, however, formalism was still little more than a series of rough ideas that were not carefully thought out or formulated, but it was enough that leadership in research into the foundations of mathematics began to pass from England to Germany, from Russell to Hilbert and followers. This created a small problem in that, as usual, there were those who would take an idea and run it beyond its intended boundaries. In its exaggerated form, formalism became a kind of caricature of what he had in mind. This was the idea that mathematics is merely a way of manipulating unlabeled or uninterpreted symbols, and that it was therefore little more than a game and had little significance. This in fact harks back, to some extent, to the position held by Thomas Henry Huxley (chapter 5), but Hilbert offered the additional factor of his "uninterpreted" geometric symbols that could be dealt with by formal rules. Hilbert would have nothing to do with this exaggerated version of his ideas.

In any case, he saw little reason to carry the formalist idea forward any further and turned his attention to mathematical analysis and especially integral equations, as well as to problems in mathematical physics. He didn't return to foundational questions until 1917–1918.

In the meantime, however, Luitzen E. J. Brouwer, a Dutch mathematician, was taking a position that stood in direct opposition to Hilbert's and, at the same time, was establishing himself as the standard-bearer of a mathematical school that came to be called intuitionism. He believed that there are built-in human thought patterns that are the basis for mathematics, and that much of what was being put forth as mathematics was mere window dressing.

As one of Brouwer's biographers, Walter van Stigt, has written, "Neither Brouwer nor Hilbert were temperamentally capable of keeping mathematical controversy at the level of a detached professional debate. Brouwer in particular needed the stimulus of a personal challenge to stir him into action; he was a fighter, who needed a personal enemy on whom to concentrate his attack. Even if Hilbert's and Brouwer's views on mathematics could hardly be described as antipodal in every respect, the foundational debate now became polarized into a battle between intuitionism and formalism with international leadership as the prize."[8]

Dirk van Dalen, a professor of mathematics and philosophy at the University of Utrecht in the Netherlands, has done a more recent, two-volume biography of Brouwer. He feels that Hilbert acted from early on as the insulted prima donna, that it was Hilbert who set the tone of the later battle. "There is no doubt that Brouwer did not tolerate plagiarism or insults," says van Dalen, "but he would not start a conflict. It was always in defense that he acted."

The result, says van Dalen, is that "The conflict that should have been a purely scholarly debate took a personal turn, when in 1928 Hilbert overstepped the lines of scholarly debate and attacked Brouwer's position in the mathematical community. This, even though Brouwer had shrugged his shoulders at the hostilities in Hilbert's earlier papers and talks." Van Dalen feels that Hilbert got away with this because of his position in the field: no one dared to criticize his papers. "When the actual break came, Hilbert clearly was angered by Brouwer's effective stonewalling, and Brouwer was not the person to tolerate injury with the added insult."[9]

Result: van Dalen describes the two men as "the chief antagonists in [what would turn into] the most prominent conflict in the mathematical world of this century."[10] Einstein, on the editorial board of a journal that played an important role in the fracas, could have been a key figure in this battle but insisted on remaining neutral. By 1928, when the drama had come to a serious and rather brutal climax, Einstein described the feud between the two men and their followers as the War of the Frogs and the Mice. Hilbert was the main mouse.

The Mouse

Although one would hardly guess it from Einstein's use of the term *mouse*, Hilbert was by the teens of the 20th century probably the greatest mathematician of his day (Poincaré had died in 1912). Born in a suburb of Königsberg in 1862, he received his early education there. His biographer Constance Reid says that his early schooling was quite traditional and would normally have depended on memorization, but "he seemed never really able to understand anything until he had worked it through in his own mind."[11] He was, however, good at mathematics, which would fit in well with that kind of mind. Reid says that a niece later recalled that the whole family saw him as a bit off his head.[12]

He entered the university of Königsberg in 1880, where he obtained his Ph.D. five years later. By 1895, he was a full professor at the University of Göttingen, which grew into one of the few important centers of mathematics not located in a major city. By then, he had already shown the depth and the breadth of both his interests and his abilities. Over the course of his career, there were few areas in which he didn't make original contributions.

Thanks at least partly to Hilbert's growing reputation, Göttingen's reputation was growing apace and attracting other potentially important mathematicians, both faculty and students. Among them was Hermann Minkowski (1864–1909), whom he recruited to the faculty, and who was to become a dear friend and colleague. It also included Zermelo as a privatdozent in 1899 and Hermann Weyl (1885–1955), who arrived in 1903 as an 18-year-old "country boy."

Weyl wrote later, while at the Institute for Advanced Study in Princeton, New Jersey, "I made bold to take the course that Hilbert announced for that term, on the notion of number and the quadrature of the circle. Most of it went straight over my head. But the doors of a new world swung open for me."[13] Weyl would be a major player in the coming battle between Hilbert and Brouwer.

Hilbert's broad interests were a factor in the controversy that eventually erupted. His underlying interests in the foundations problem were one factor, while his earlier work *The Foundations of Geometry* led him back to this field of mathematics. As I showed in the last chapter, the book was his attempt to recast Euclid in a rigorous axiomatic fashion by use of the principles of Peano, though without Peano's complex symbolic system. One of Peano's objectives had been to put mathematics into a formal language without any need to call upon intuition, which he distrusted. He also felt that using Euclid as the model or the basis for geometry was a mistake. He, and others by this time, felt that although the structure of Euclid's geometry was deductive, it was full of hidden assumptions, poor definitions, and mistakes in logic. Hilbert hoped to establish a more solid foundation for geometry and at the same time eliminate any dependence on intuition.   *unalterable?*

Other developments in geometry had also caught his interest. Earlier in the century, Nikolai Lobachevsky, Janos Bolyai, Bernhard Riemann, and Carl Gauss had shown that geometries other than Euclid's were possible, with the result that one of the main postulates of Euclid (re parallel lines) was no longer accurate. Furthermore, in Euclidean geometry the sum of all the angles of a triangle is 180 degrees; in the non-Euclidean geometries other sums are possible. Hilbert felt that in order to tie geometry, and hence mathematics, down more explicitly, something had to be done by way of eliminating certain assumptions.

By siding with Cantor, he was already involved in controversy. He also felt that Kronecker's beliefs threatened the progress of mathematics. Kronecker, you'll recall, wanted to keep the subject bound by a highly subjective intuitive foundation, one that called for step-by-step construction and restriction to a real, material world. Hilbert, in contrast, asked only for logical progression and consistency.

Irrational numbers, to which Kronecker objected so strongly, should not be kept out of the world of numbers, he was convinced. Without them, the world of analysis would be condemned to sterility.

Furthermore, Hilbert felt that in order to strengthen the foundations of mathematics, it would be necessary to support Cantor's ideas on the infinite. He saw this in fact as more than a mathematical issue. He wrote, "The definitive clarification of the nature of the infinite has become necessary, not merely for the special interests of the individual sciences but for the honor of human understanding itself."[14]

Stephen G. Simpson, however, writing in the *Journal of Symbolic Logic*, argues that Hilbert's idea of reconstituting infinitistic mathematics into a big, elaborate formal system "led to an unnecessary intellectual disaster." Simpson says, "It left Hilbert wide open to Brouwer's accusation of 'empty formalism.'"[15] Here's how it happened.

The Frog

Born on February 27, 1881, in Overschie, now a suburb of Rotterdam, Netherlands, Luitzen Egbertus Jan Brouwer is often known simply as L. E. J. Brouwer. He did well in school and while still an undergraduate at the University of Amsterdam did original work on continuous motions in four-dimensional space. This led to a publication by the Royal Academy of Science in Amsterdam while he was still an undergraduate. From 1904 to 1907, he did studies in philosophy and mysticism, and in 1905, he wrote a book titled *Life, Art and Mysticism*. His ideas reflected some of the romantic thinking of the time and came out in a rejection of the domination of nature (that is, the world) by mankind. This included industrial exploitation and subjugation of the environment. As a true mystic, he denied the possibility of accurate communication and the role of language. These ideas would color his feelings about formalism.

What kind of man was he? There is no question of his brilliance. As for his character, expressions like misanthropic, self-centered, high strung, emotional, and stubborn are used. In van Dalen's biography of Brouwer, he writes, "Brouwer was a high-strung nervous person, who could easily exaggerate matters when under stress. On top of that he had an extreme passion for justice; as Ludwig Bieberbach [whom we'll meet later] put it: he was a justice fanatic (*Gerechtigkeitsfanatiker*). As a result he would counter injustice—no matter with respect to whom—with a state of total war."[16]

As a student, he was withdrawn and difficult in his social contacts. Under the guidance of his best friend (a prominent socialist poet), he learned to move in society. Gradually, he became an inveterate talker with a hunger for company.

By 1907, he had his doctorate in mathematics and physics from the University of Amsterdam. His interests there spread out to topology and the foundations of mathematics, both of which he would make major contributions to over his lifetime. Though still just a graduate student, he had some strong ideas on the existing discussions in mathematics, which he offered in his doctoral thesis *On the Foundations of Mathematics*.

Regarding the Poincaré/Russell fracas, it was clear that he came down on Poincaré's side; he argued that although logicism might be useful in certain circumstances, it could not provide a solid foundation for mathematics. Whereas Russell was claiming that mathematics depends on logic, Brouwer maintained that logic depended on mathematics. In addition, although Hilbert had not yet fully developed his ideas on formalism, there was enough there for Brouwer to know he did not like them. He criticized Hilbert's program, arguing for instance that there was no guarantee of an acceptable mathematical structure that could satisfy the call for a consistent mathematical theory. He was also critical of Cantor's theory of transfinite numbers.

It's interesting that in spite of Brouwer's strongly negative comments on Hilbert's formalist program, this did not seem to have any effect on Hilbert's feelings toward Brouwer, no doubt because the thesis was written in Dutch and was not widely circulated. They met, for example, in 1909 at Scheveningen, a fashionable seaside resort. The meeting apparently went well. Brouwer, 19 years younger than Hilbert, explained his language and mathematics levels to Hilbert and later described Hilbert in a letter to a friend as "the first mathematician in the world." Hilbert not only recommended Brouwer for a professorship in Amsterdam in 1912 (Brouwer had been an unsalaried lecturer until then), but also offered him a professorship in Göttingen—a definite step up—as late as 1919. Brouwer refused (which may have been a factor in the coming breakup). Brouwer remained at Amsterdam until his retirement in 1951.

Brouwer's work on topology and his ideas on foundational mathematics earned him a solid reputation, in spite of the fact that his ideas on intuitionism were not widely accepted in the early years. In 1912, he was elected to membership in the Royal Netherlands Academy of Science. After that, the Prussian Academy of Science in Berlin, the American Philosophical Society, and the Royal Society in London all elected him a member. He also received several honorary doctorates.

Though he did other work, his main concern was with the foundations question. In 1908, he produced a treatise titled "On the Unreliability of the Logical Principles," which rejected as invalid mathematical proofs that had been produced using one of the cardinal laws of logic, a common technique called the principle of the excluded middle (PEM). The principle of the excluded middle asserts that every mathematical statement is either true or false and no other possibility is allowed. Then, in 1912, in his inaugural address as professor of mathematics at the university, he further considered problems that he saw associated with this "law."

Brouwer saw PEM as an example of a logical principle that is applied too freely. The principle asserts that every meaningful statement is either true or false and is basic to the so-called indirect method of proof, which permits use of the conventional logical reductio ad absurdum, or proof by contradiction. Here one can prove something is true by proving that a logical contradiction arises if it isn't true. Brouwer denied the PEM and insisted on the existence of a third level, "undecided," for statements whose truth or falseness had not been constructed via a finite number of deductive steps. He invariably challenged mathematical proofs that were based on the PEM: he described them as "so-called proofs."[17]

He stated in 1920, "The use of the Principle of the Excluded Middle is *not permissible* as part of a mathematical proof. . . . [It] has only scholastic and heuristic value, so that theorems which in their proof cannot avoid the use of this principle lack all mathematical content."[17]

Hilbert answered, "Taking the Principle of the Excluded Middle from the mathematician is the same as . . . prohibiting the boxer the use of his fists."[18] It was in truth an extremely limiting requirement and was found hard to accept by many mathematicians and scientists, especially those who made wide use of the procedure.

In the years following, Brouwer began to campaign for his views, which began to take over his life. Though in the years 1909 and 1913, he had done some excellent work in the field of topology, he never gave lectures on topology. Bartel L. van der Waerden, who took a course with him, reported that he never looked at the students, wanted no questions, and always lectured on the foundations of intuitionism.

Interviewed later, van der Waerden said, "It seemed that he was no longer convinced of his results in topology because they were not correct from the point of view of intuitionism, and he judged everything he had done before, his greatest output, false according to his philosophy. He was a very strange person, crazy in love with his philosophy."[19]

Brouwer, of course, saw things a little differently. In his 1919 paper "Intuitionistic Set Theory," Brouwer pointed out that his earlier topological work had not been intuitionistically correct; he added that most of it could be salvaged in an intuitionistic framework. He showed in a few examples (for example, the fixed point theorem) how an intuitionistic version could be proved.[20]

In the years 1917–1920, Brouwer had begun to develop his intuitionistic ideas further, including development of set theory along intuitionistic lines. After about 1920, he also decided it was time to reach out to the outer world. This was a clear challenge to Hilbert's work. Then, to make matters worse, Weyl, who had published an alternative foundation of analysis called *The Continuum*, switched sides at this time. In 1921, he even published an article explaining his new position, saying in part that Hilbert's approach reduces everything to a kind of game. Weyl had been Hilbert's star student, and Hilbert had counted on him as one of his devoted followers. In the spring of 1920, Brouwer wrote up some comments on a manuscript that Weyl had generated, spelling out his ideas, and sent it to Weyl. Brouwer began, "Your unreserved support has been a source of infinite joy. . . . The fact that we disagree on some minor points can only have a stimulating effect on the reader."[21]

Hilbert took this turnabout very hard. Ironically, Weyl, after about 1925, attempted to take a middle ground between the two contenders, but by then Hilbert had already set his sights.

Counterattack(s)

In 1922, when Hilbert began his counterattack, it was by an article aimed at both Weyl and Brouwer. He first stated his belief that a deeper treatment of the foundations of mathematics is required than had been accomplished to that point. He then wrote, "Distinguished and highly accomplished mathematicians, Weyl and Brouwer, are seeking the solution to these problems by following what I believe to be a false path."[22] He then explained why and went on to lay out his new ideas.

He was careful to point out that his approach to the grounding of the continuum "is not at all opposed to intuition. The concept of extensive magnitude, as we derive it from intuition, is independent of the concept of number, and it is therefore thoroughly in keeping with intuition if we make a fundamental distinction between number and mass-number or quantity." This matter has been studied carefully by others, he wrote, then added, "If Weyl here sees an 'inner instability of the foundations on which the empire is constructed,' and if he worries about 'the impending dissolution of the commonwealth of analysis,' then he is seeing ghosts."

Hilbert continued:

> To be sure, the problem arises of proving the consistency of the axioms; this is a well-known problem, and for decades I have never lost sight of it. This report concerns the solution of this problem.
>
> What Weyl and Brouwer do amounts in principle to following the erstwhile path of Kronecker: they seek to ground mathematics by throwing overboard all phenomena that make them uneasy and by establishing a dictatorship of prohibitions *à la*

Kronecker. But this means to dismember and mutilate our science, and if we follow such reformers, we run the danger of losing a large number of our most valuable treasures. . . . I believe that, just as Kronecker in his day was unable to get rid of the irrational numbers . . . so today Weyl and Brouwer will be unable to push their programme through. No: Brouwer is not, as Weyl believes, the revolution, but only a repetition, with the old tools, of an attempted coup that, in its day, was undertaken with more dash, but nevertheless failed completely; and now that the power of the state has been armed and strengthened by Frege, Dedekind, and Cantor, this coup is doomed to fail.

After some more introductory material, he argued that "there has scarcely been a serious attempt to represent the consistency of the axioms, whether in number theory or analysis or set theory."[23]

Here, in answer to the attacks of Weyl and Brouwer, was the basis for his revised and bold new program, which set a new and more solid foundation for the school called formalism. The article went on to give his idea for an outline of a proof of the consistency of the axioms of analysis, building number theory on the basis of the signs 1 and +. "When we develop number theory in this way, there are no axioms, and no contradictions of any sort are possible."[24] He then rounded out the rest of the report. Dirk van Dalen maintains, however, that "Hilbert never succeeded in getting his proof right. It was something like a bulge in the tube of a bicycle tyre [sic]; each time he pushed in the bulge, another one appeared somewhere else. To be fair, the matter was terribly complicated. Years later, in 1936, Gerhard Gentzen, a brilliant German mathematician (1909–1945), finally managed a complete and correct analysis of the transfinite process."[25] With Gentzen's update, Hilbert's proof theory has been demonstrated to be an elegant and powerful mathematical tool, and of special interest in computer science.[26]

There is an ironic postscript to the Weyl story. In this same 1922 article, Hilbert attacked both Brouwer and Weyl as a team, ignoring some clear differences in their ideas, as, for example, their interpretation of intuitionism. Then, as I noted earlier, around 1925 Weyl began to move away somewhat from Brouwer's intuitionism, and from then on Hilbert directed his attacks mainly at Brouwer, while ignoring Weyl's change of heart. Even as late as 1928, for example, Hilbert brought up and answered the criticism that his school reduces everything to a game. You'll recall, however, that this was Weyl's criticism, not Brouwer's.

It's important to understand that both men by this time saw their feud as existing on two levels. There was, of course, the personal competition, but both also saw themselves as saviors in a broader sense. That is, both saw mathematics in a foundational crisis, and each saw himself as the one who could save mathematics from decline and destruction. In addition, Hilbert's fears, at least, went even deeper. He felt, "If mathematics fails, so does the human spirit."[27]

In one of Brouwer's responses to Hilbert's new approach, he wrote in 1928, "All this shows that Formalism has received nothing but favors from Intuitionism and can expect further benefits. The Formalist school, therefore, should give some recognition to Intuitionism instead of attacking it in sneering tones and not even making proper mention of authorship. Moreover, Formalism should consider the fact that within the framework of Formalism so far *nothing* of mathematics proper has yet been secured (since a proof of consistency of the axiom system is still missing).

"On the other hand, Intuitionism . . . has already built a new structure of mathematics proper with unshakeable certainty."[28] As far as Brouwer was concerned, any language, including the formalistic one, might be useful but only for communication. As one of Brouwer's doctoral students, Arend Heyting, put it nicely in 1930: "It is in principle impossible to set up a system of formulas which would be equivalent to intuitionistic mathematics, for the possibilities of thought cannot be reduced to a finite number of rules set up in advance."[29]

In retrospect, it can be seen that Hilbert and Brouwer were talking at cross purposes, and at least at one point their difficulties involved a clash of national feelings—with, oddly, Brouwer taking up the cause of German mathematicians: for a Riemann memorial volume, he objected strongly to the inclusion of some French mathematicians, much to Hilbert's annoyance. Finally, their arguments were taking on a more personal bias—on both sides.

A Ruthless Attack

Brouwer, though not as well known as Hilbert, was nevertheless building a solid reputation in the mathematical world. He began to publish papers on his philosophy in such journals as the *Mathematische Annalen* of 1925 and 1926. He had been appointed to the editorial board of this respected journal in 1914 and had served carefully—and slowly—since then. Such an appointment was an honor and a sign of the respect in which he was held in the mathematical community.

Yet his irascibility would have its effect even here. One of the main editors had been Felix Klein. Klein had chosen to resign, apparently over a dispute with Brouwer—in a matter in which Brouwer was actually in the right but acted so rudely that Klein decided to resign. This was an astonishing outcome, for Klein had long been associated with the *Annalen*, had become a sort of editor in chief, and had had a major hand in its success.

Yet Hilbert's name was also on the masthead of the *Annalen*. Unfortunately for Brouwer, who might today be called an associate editor, Hilbert was one of the chief editors. In fact, after Klein's resignation, Hilbert had taken over many of his duties, with the result that the journal began to take on the reputation of being "owned" by Göttingen mathematicians. The level to which the feud had escalated can be judged from Hilbert's next move.

The Frog and Mouse War

Shortly before the end of October 1928, Brouwer was given a letter written by Hilbert. It stated,

Dear Colleague:

Because it is not possible for me to cooperate with you, given the incompatibility of our views on fundamental matters, I have asked the members of the board of managing editors of the *Mathematische Annalen* for the authorization, which was given to me by

Blumenthal and Carathéodory [Otto and Constantin, two of the other managing editors; Albert Einstein was the fourth], to inform you that henceforth we will forgo your cooperation in the editing of the *Annalen* and thus delete your name from the title page. And at the same time I thank you in the name of the editors of the *Annalen* for your past activities in the interest of our journal.

Respectfully yours

D. Hilbert[30]

The reasons for his action are more complicated than they might seem at first. The main one seems to be that Hilbert had some fears of his own early demise—he had been diagnosed with pernicious anemia in November 1925—and wanted to make very sure that Brouwer's ideas did not continue to have an inside track on the *Annalen*. There was also a personal angle. Dirk van Dalen describes Blumenthal's view of the situation: "Hilbert saw in Brouwer a headstrong, unpredictable and domineering character. He . . . feared that, when he at some time should have left the board, Brouwer would bend it to his will."[31]

Another reason had to do with the relations between German and French mathematicians, which were still strained as a result of World War I. Hilbert felt that earlier opportunities to apply some balm to these relations had been sabotaged by Brouwer. For example, plans had been made for publication of a Festschrift honoring Bernhard Riemann in 1926—that is, a volume commemorating the anniversary of his birth a hundred years earlier. Hilbert had wanted to include a paper by an important French mathematician, Paul Painlevé. In 1918, Painlevé had, however, fiercely denounced the German scientific community. Hilbert and others felt that Painlevé had later been able to move beyond such feelings, but Brouwer felt strongly that including him would insult the German mathematical establishment. In spite of Hilbert's more senior position, the essentially democratic setup of the journal's operation led to the volume being issued without any French contributors.

Similarly, in the present case Hilbert needed acquiescence from the other managing board members (*Herausgebern*). He was able to get this to varying degrees from Blumenthal and Carathéodory, but with Einstein he had hit a wall. He had written to Einstein on October 15, asking for his permission as one of the *Herausgebern* to send a letter of dismissal to Brouwer. Among the reasons he gave was that Brouwer in an earlier circular letter had insulted not only him (Hilbert) but the majority of German mathematicians; that Brouwer held a "strikingly hostile position vis-à-vis sympathetic foreign mathematicians; and that he [Hilbert] thought it would be a good idea to keep Göttingen as the chief base of the *Annalen*." He also mentioned his poor health in a postscript.

Einstein answered, in essence: do what you have to do, but I cannot sign such a letter.

In response to a letter from Carathéodory asking his advice, Einstein answered (October 19), "It would be best to ignore this Brouwer-affair. I would not have thought that Hilbert was prone to such emotional outbursts."[32] As I noted earlier, Einstein refused to go along with the scheme.

If Hilbert thought Brouwer would beat a meek retreat, though, he was mistaken. When Hilbert finally did manage to get the letter written, Brouwer took this abrupt dismissal as a direct and major insult, and he responded quickly, though not immediately. According to one report, he was subject to "nervous fits" and was ill and feverish for several days after the initial shock.

Then, however, he wrote to Carathéodory:

Dear Colleague:

After close consideration and extensive consultation, I have to take the position that the request from you to me [Carathéodory had delivered the initial letter of dismissal], to behave with respect to Hilbert as to one of unsound mind, qualifies for compliance only if it should reach me in writing from Mrs. Hilbert and Hilbert's physician.

Yours,

L. E. J. Brouwer[33]

This was a serious error on Brouwer's part. In a parlor game of some sort, it might have been taken as a clever retort. In this situation, it was seen, as it was by Blumenthal, as "this frightful and repulsive letter." He and Carathéodory began to wonder if Hilbert had perhaps "known and judged him [Brouwer] better than we did."[34]

Brouwer was unrelenting. He next wrote to Mrs. Hilbert, asking her to use her influence on her husband to change his mind. He sent a copy to Richard Courant, one of the other board members, who actually visited Hilbert's wife, but then wrote back to Brouwer telling him that Hilbert was not to be influenced by anyone.

On the same day (November 5), Brouwer appealed in a broadside to all the editors. It included several explanations that spelled out his view of Hilbert's "increasing anger against me": that Hilbert's statement about lack of cooperation was a smokescreen, for they had not exchanged any letters for years; that no objection of any sort had until then been lodged against him; that it was unfair for the managing editors to put Hilbert's state of health above his (Brouwer's) rights and honor; and that the prestige and the scientific contents of the *Annalen* were being sacrificed.

By now, Hilbert had more or less withdrawn from the fray, but the editors, except for Einstein, had pretty much taken sides, with most of them on Hilbert's side. Brouwer, as it turned out, was one of only three editors who were not German and was therefore something of an outsider. This might have been a factor in the outcome. His letter to Carathéodory re Hilbert's "unsound mind," though, most likely sealed his fate.

Brouwer, along with Professor Ludwig Bieberbach, one of the board *unter* editors, then traveled to see the publisher, Ferdinand Springer, in Berlin. They both made threats against the *Annalen* and the Springer interests if the dismissal was carried out. Brouwer threatened to found a competing journal, but Springer would not be intimidated.[35]

Blumenthal, originally torn but now solidly in Hilbert's camp, also circulated a letter to the entire board, answering Brouwer's charges. One of the points he made: "True, Brouwer was a very conscientious and active editor, but he was quite difficult in his dealings with the managing editor and he subjected the authors to hardships that were hard to bear.

"For example, manuscripts that were submitted to him for refereeing lay around for months."[36] Dirk van Dalen points out, however, "What Blumenthal did not mention was that in the past he had, in his capacity as managing editor, often used Brouwer as a trouble shooter, and that he had never complained to Brouwer about his handling of the manuscripts."[37]

The deed was not yet done, though, and a key figure remained to be persuaded. If Einstein could be convinced to join the Hilbert side, it would be easier to eliminate the remaining opposition. Einstein's colleague Max Born tried to win over Einstein personally. Responding in a letter of November 27, Einstein clearly stated his strict neutrality, and it was here that he introduced his apt characterization of the situation as a *Frosch-Mäusekrieg* (War of the Frogs and the Mice).[38] He wrote to Brouwer and Blumenthal, "I am sorry that I got into this mathematical wolf-pack like an innocent lamb. . . . Please allow me[,] therefore, to persist in my 'booh-nor-bah' (*Muh-noch-Mäh*) position and allow me to stick to my role of astounded contemporary."[39]

There was more. Lawyers were called in. Brouwer sent a letter to the editors (January 23, 1929), in which he accused Hilbert and Blumenthal of "embezzlement" at the *Annalen*. Dirk van Dalen explains that Brouwer was here using the term in a metaphorical sense. Brouwer felt that the *Annalen* were given in trust to the editors in chief by the (German) mathematical community, and that by Hilbert's action, that trust was violated.[40] After that, Brouwer offered a final shot: a long letter summarizing his grounds for why he should not have been dismissed, including an alternate explanation of the Klein affair. None of this helped him.

Finally, it was done. The solution was to dissolve the old editorial board and form a new one—but with a crucial difference. As shown on the new cover for 1929, there would only be *Herausgebern* and no *Mitarbeiteren*, at least none shown on the cover. Hilbert's name remained on the cover. This way, it would appear to be just a major change in policy, rather than an act against one of the editors, that is, Brouwer.

The Winner and New Champion

Quite clearly, Hilbert had won. Dirk van Dalen, who has studied the *Frosch-Mäusekrieg* carefully, calls the whole affair a tragedy of errors. Although he feels that Hilbert's annoyance with Brouwer was understandable, "Hilbert's illness, with the real danger of a fatal outcome, must have influenced his power of judgment. . . . One has to agree with Einstein: if Brouwer was a menace of some sort, there were other ways to safeguard the *Annalen*. . . . Most likely the letter to Einstein shows an unguarded Hilbert with personal motives after all."[41]

As for his fears that Brouwer would, had he been given the chance, have turned the *Annalen* into a bastion of intuitionism, it is interesting to note that not even Brouwer's own journal, the *Compositio Mathematica*, was made to function in that fashion.

Though Brouwer lived, traveled, and lectured for another 36 years (he died in an automobile accident in 1966 at the age of 85), it was mainly in other areas of mathematics. He retreated into a shell as far as the foundations question was concerned. The lack of backing from people he had considered colleagues, the rather ruthless dismissal from the *Annalen*, and his own sometimes shaky psychological makeup drove him into what van Dalen calls a self-chosen isolation.

He certainly looked upon Hilbert as "my enemy." At one point, he walked out of a gathering when van der Waerden, who was also a guest, referred to Hilbert and Courant as his (van der Waerden's) friends.[42]

Brouwer worked a bit more with his intuitionism, along with a few followers, but the excitement was gone, and although intuitionism as a school was not likely to have become a dominant force, it might have had a greater impact if he had been able to continue his crusade.

There was, however, an ironic long-term outcome to the battle. The immediate result was that formalism, as a contender, looked pretty good—for a while. Hilbert's victory didn't last long, though, for in 1930–1931, the young Austrian logician Kurt Gödel came up with a proof that in essence showed that Hilbert's formalism program could not really be carried out. (See chapter 7.) Gödel's work shook the mathematical world and in truth also blasted Hilbert's optimistic belief in the solvability of all mathematical problems.

Yet Hilbert's iron-willed optimism would not let him believe this. In 1931, his last recorded words over the radio stated, "*Wir mussen wissen. Wir werden wissen.*" (We must know. We shall know.) The words are also inscribed on the stone over his grave. He died in 1943.

The irony I mentioned earlier is that Gödel's paper was most damaging to the formalists, so that only the intuitionists who were still active could still hold up their heads and say I told you so. In general, however, mathematicians tended to turn away from the foundations question, away from the philosophy of mathematics.

"Nevertheless," writes Ernst Snapper in an essay in *Mathematics Magazine*,

> the influence of the three schools [logicism, formalism, intuitionism] has remained strong, since they have given us much new and beautiful mathematics. This mathematics concerns mainly set theory, intuitionism and its various constructivist[43] modifications, and mathematical logic with its many offshoots [including foundational work for the computer]. However, although this kind of mathematics is often referred to as "foundations of mathematics," one cannot claim to be advancing the philosophy of mathematics just because one is working in one of these areas. Modern mathematical logic, set theory, and intuitionism with its modifications are nowadays technical branches of mathematics just as algebra or analysis, and unless we return directly to the philosophy of mathematics, we cannot expect to find a firm foundation for our science. It is evident that such a foundation is not necessary for technical mathematical research, but there are still those among us who yearn for it.[44]

Indeed, the grand, unified theory of mathematics remains elusive. Snapper believes "that the key to the foundations of mathematics lies hidden somewhere among the philosophical roots of logicism, intuitionism, and formalism."[45]

10

Absolutists/Platonists versus Fallibilists/Constructivists

Are Mathematical Advances Discoveries or Inventions?

In a world filled with doubt and uncertainty, mathematics has long been seen as the last fortress of certitude. Holders of a view variously called absolutism or Platonism reflect this idea and see mathematics as objective and precise. They cite its remarkable ability to describe activities and patterns in both nature and technology, and argue that true mathematical knowledge is perfect and eternal.

In opposition are mathematicians who also go under a variety of names, one of the most common being fallibilists (suggesting mathematical fallibility). They see mathematics as a work in progress. Some even argue that certain advances are accepted on the basis of the mathematician's authority and not on rational proof.

Another group of mathematicians who would fall under this category are the constructivists. Their ideas can be traced back to Kant and Kronecker (chapter 6). The constructivist objective is to restructure mathematical knowledge to keep it from obsolescence and contradictions. Thus constructivists reject Cantor's proof that the real numbers are uncountable, and they reject the law of the excluded middle. The intuitionist L. E. J. Brouwer (chapter 9) fell into this category. In other words, certain parts of classical mathematics are unsafe and must be "reconstructed" by "constructive" thinking and methods.

Curiously, science has gone through a somewhat similar evolution. With the advent of relativity and quantum mechanics, absolutist views in science have tended to give way to fallibilistic thinking. In mathematics, however, a strong core of absolutism/Platonism remains intact—in fact, may even be the dominant mode—although it is under increasing attack from a number of sides.

Paul Ernest, editor of the online journal *Philosophy of Mathematics Education*, states that "in the past few decades a new wave of 'fallibilist' philosophies of mathematics has been gaining ground, and these propose a different and opposing image of mathematics as human, corrigible, historical and changing. Fallibilism views mathematics as the outcome of social processes. Mathematical knowledge is understood to be eternally open to revision, both in terms of its proofs and its concepts. Consequently this view embraces the practices of mathematicians, its history and applications, the place of mathematics in human culture, including issues of values and education as legitimate philosophical concerns."[1]

The philosophical angle is not a new one. René Thom, a well-known French mathematician with a strong interest in the philosophy of the subject, wrote in 1990, "The philosophy of mathematics is in the midst of what might be termed a 'Kuhnian revolution.'"[2] The term *Kuhnian* here has several implications.

- It denotes a significant turnabout in the way things are done. For example, in the 1960s, set-theoretic notation and axiomatics were incorporated into the high school mathematics curriculum. How's that for a starter?

- It implies that the revolution is an advance and not merely a change. As we will see, there is some question about this, but Thom felt that the term *Kuhnian* is still appropriate.

- In addition, the activity takes place in an important area of interest. That is certainly true.

Thom saw two major reasons why absolutism is giving way in mathematics. "One reason for this," he wrote, "is that the foundations of mathematics are not as secure as was supposed. Gödel's first Incompleteness theorem has shown that axiomatics must fail to capture the truths of most interesting mathematical systems [see chapter 7].

"Another reason is a growing dissatisfaction amongst mathematicians, philosophers and educators with the traditional narrow focus of the philosophy of mathematics, limited to the foundations of pure mathematical knowledge and the existence of mathematical objects."[3] There is, in other words, a growing feeling in the world of mathematics that all of its branches—research, philosophy, history, teaching, and learning—are connected, and that in all of those areas absolutist thinking is sterile and restrictive.

So there we have it: a strong force of reformers, drawn from the ranks of mathematicians, philosophers, and educators, who see mathematics as fallible, corrigible, and open to correction and revision, battling, as we'll see, against an equally strong, equally diverse cadre who still hold to the original idea of mathematics as the last bastion of certainty.

Yet these two modes of thinking go hand in hand with another strongly contested division of thought: is mathematics discovered or created? After all, if true mathematical knowledge is perfect and eternal, then whatever new ideas mathematicians come up with must be discoveries. But if mathematics is fallible, a work in progress, then new mathematical ideas must be created.

Or, as Morris Kline puts the basic question: "Is then mathematics a collection of diamonds hidden in the depths of the universe and gradually unearthed, or is it a collection of synthetic stones manufactured by man, yet so brilliant nevertheless that they bedazzle those mathematicians who are already partially blinded by pride in their own creations?"[4]

This, then, is our basic question. Let's discuss this first, then see where it takes us.

A Collection of Diamonds to Be Discovered

The list of people who have thought that mathematics is a collection of diamonds to be discovered is long and impressive. Cantor, you'll remember, believed that he was merely a reporter, that set theory and the infinite had been revealed to him by God (chapter 6).

The list starts much earlier, though. Plato was one of the first to stand in this corner. Basically, he argued that there are two worlds. There is the concrete world, that is, the world of real things, which we perceive through our senses; and then there is the abstract world—the world of the spirit, that is, the world of ideas and concepts, such as goodness, justice, beauty, and perfection. The circles, the squares, and the parallel lines that we draw are imperfect; they belong to the concrete world. Yet somewhere there exist perfect ones, of which we can only conceive. They are ideal, unchanging, eternal. They would be there even if we did not exist. The same concept applies to the numbers and to the mathematical functions. In short, we discover the mathematical truths, we do not invent them.

Among the better known current exponents of the absolutist viewpoint is John D. Barrow, a British author and professor of astronomy at the University of Sussex. Barrow, for reasons of his own, prefers the term *Platonist*. He writes, "The existence of mathematical entities inhabiting some realm of abstract ideas is a lot for many modern mathematicians to swallow, but three hundred years ago a Newton or a Leibniz would have taken for granted the existence of mathematical truths independent of the human mind. They had faith in the existence of the Divine Mind in which perfection lived and so they saw no problem at all with the concept of perfect forms. Their problem was to reconcile them with the existence of the imperfect, material objects they saw around them."[5]

We are reminded of Newton's famous comment in which he pictured himself as a boy playing on the seashore, picking out a pebble or a prettier shell than the ordinary, while the great ocean of truth lay undiscovered before him.

Charles Hermite, a respected French mathematician who died in 1901, expressed a similar idea: "I believe that the numbers and functions of analysis are not the arbitrary products of our spirits; I believe that they exist outside of us with the same character of necessity as the objects of objective reality; and we find or discover them and study them as do the physicists, chemists, and zoologists."[6]

The great British analyst G. H. Hardy wrote in 1929, "It seems to me that no philosophy can possibly be sympathetic to a mathematician [who] does not admit, in one manner or other, the immutability and unconditional validity of mathematical truth. Mathematical theorems are true or false; their truth or falsity is absolutely independent of our knowledge of them. In *some* sense, mathematical truth is part of objective reality."[7]

We grant that these quotes were written before the appearance of Gödel's theorem (1930–1931), but even after its publication, the absolutist idea has remained a strong one.

Hardy, for example, continued to express the same absolutist view. Thus, in his later book *A Mathematician's Apology* (1941), he wrote, "I believe that mathematical reality lies outside us, that our function is to discover or observe it, and that the theorems which we prove, and which we describe grandiloquently as our 'creations,' are simply our notes of our observations."[8]

In 1945, Jacques Hadamard, a leading French mathematician (chapter 7), maintained (in his *Psychology of Invention in the Mathematical Field*), "Although the truth is not yet known to us, it *pre-exists*, and inescapably imposes on us the path we must follow."[9]

You may have noted in the words of Morris Kline, which ended the previous section, that he refers to discovered gems as diamonds, while the ones that are man-made are "so brilliant . . . that they bedazzle those mathematicians who are already partially blinded by pride in their own creations." It seems clear that as far as he's concerned, the *discovered* gems are the real diamonds. Kline's description was written in 1980.

As we saw earlier, writers who hold these views are sometimes called Platonists. Kline believed, however, that there's a problem with this term. He said that Plato did indeed believe that mathematics exists in a kind of ideal world apart from human beings, but that his doctrines no longer apply in today's world. As a result, he argued, "the use of the appellation Platonist is more unsuitable than helpful."[10]

Barrow obviously disagrees. He points out that, in fact, "this philosophy of mathematics only became generally known as 'mathematical Platonism' after 1934, when it was so described by Paul Bernays, Hilbert's close collaborator on the development of consistency proofs for formal mathematical systems."[11] In any case, both terms, *absolutist* and *Platonist*, are widely used.

Barrow sums up this side of the basic controversy: "The Platonic view of reality has crept unseen upon many modern scientists and mathematicians. It seems simple, straightforward, and inspiring. There is an ocean of mathematical truth lying undiscovered around us; we explore it, discovering new parts of its limitless territory. This expanse of mathematical truth exists independently of mathematicians. It would exist even if there were no mathematicians at all—and indeed, once it did, and one day perhaps it will do so again. Mathematics consists of a body of discoveries about an independent reality made up of things like numbers, sets, shapes, and so forth."[12]

He adds an interesting idea: "If our minds have derived a special mathematical facility from the real world, it is likely that they have done so as a result of an evolutionary process which has selected for those mental images and representations of the world because they most faithfully represent how the world truly is."[13]

Perhaps the best-known example of the absolutist/Platonist point of view is Kurt Gödel. He refers to the entities with which the logician and the set theorist work: "Despite their remoteness from sense experience, we do have something like a perception also of the objects of set theory, as is seen from the fact that the axioms *force themselves upon us as being true*"[14] (my italics).

Finally, says Barrow, Platonism "fails to provide insight into the fact that Nature is best described by our mental inventions in those areas furthest divorced from everyday life and those events that directly influence our evolutionary history. In the end, one cannot help but feel that humanity is not really clever enough to have 'invented' mathematics."[15]

Mathematical Knowledge Is Created

Those on the opposing side, who believe that mathematical advances are created, obviously disagree. This side boasts an equally distinguished cast of advocates.

A good example is the brilliant 18th-century German philosopher Immanuel Kant, who saw the wellspring of new mathematics in the subtle workings of the mind. Our minds, he suggested, have built into them the forms of space and time. He called these forms intuitions. Space and time are filters through which our minds view the world, and this helps us to comprehend and organize the sensations that constantly bombard us. The development of mathematics runs parallel to the progressive development of the mind itself. The axioms and the theorems of mathematics are a priori synthetic judgments, to distinguish them from analytic/sense-based experiences. As a result of these ideas, some writers use the term *Kantianism* when referring to fallibilism.

In Poincaré's famous essay "Mathematical Creation" (1908), he asked, "What is mathematical creation?" and answered,

> It does not consist in making new combinations with mathematical entities already known. Anyone could do that, but the combinations so made would be infinite in number and most of them absolutely without interest. To create consists precisely in not making useless combinations and in making those which are useful and which are only a small minority. Invention is discernment, choice. . . .
>
> The mathematical facts worthy of being studied are those which, by their analogy with other facts, are capable of leading us to the knowledge of a mathematical law just as experimental

facts lead us to the knowledge of a physical law. They are those which reveal to us unsuspected kinship between other facts, long known, but wrongly believed to be strangers to one another.

In another section, he wrote, "Most striking at first is this appearance of sudden illumination, a manifest sign of long, unconscious prior work. The role of this unconscious work in mathematical invention appears to me incontestable."[16]

Joseph Dauben maintains, "Georg Cantor basically invented transfinite set theory when he discovered there were certain relations between point sets that he could generalize to resolve very difficult matters concerning trigonometric series and the nature of the set of all natural numbers."[17]

David Hilbert had a perfect explanation for why a certain mathematician had become a novelist. "But that is completely simple," Hilbert explained. "He did not have enough imagination for mathematics, but he had enough for novels."[18]

The distinguished American physicist Percy W. Bridgman argued in 1927, "It is the merest truism, evident at once to unsophisticated observation, that mathematics is a human invention."[19]

Edward Kasner and James Newman maintained in 1940,

The first significant appraisal of mathematics was occasioned only recently by the advent of non-Euclidean and four-dimensional geometry. That is not to say that the advances made by the calculus, the theory of probability, the arithmetic of the infinite, topology . . . are to be minimized. Each one has widened mathematics and deepened its meaning as well as our comprehension of the physical universe. Yet none has contributed to mathematical introspection, to the knowledge of the relation of the parts of mathematics to one another and to the whole as much as the non-Euclidean heresies.

As a result of the valiantly critical spirit which engendered the heresies, we have overcome the notion that mathematical truths have an existence independent and apart from our own minds. It is even strange to us that such a notion could ever have existed.[20]

The Hungarian-born philosopher of mathematics Imre Lakatos expounded further on the non-Euclidean angle as a factor in his list of objections regarding infallibilist (his term for absolutist) thinking. In his widely acclaimed book *Proofs and Refutations* (1976), he explained:

> It was the infallibilist philosophical background of Euclidean method that bred the authoritarian traditional patterns in mathematics, that prevented publication and discussion of conjectures, that made impossible the rise of mathematical criticism. Literary criticism can exist because we can appreciate a poem without considering it to be perfect; [but] we only appreciate a mathematical or scientific result if it yields perfect truth. A proof is a proof only if it proves; and it either proves or it does not. The idea . . . that a proof can be respectable without being flawless, was a revolutionary one in 1847, and, unfortunately, still sounds revolutionary today.
>
> It is no coincidence that the discovery of the methods of proofs and refutations occurred in the 1840s, when the breakdown of Newtonian optics (through the work of Fresnel in the 1810s and 1820s), and the discovery of non-Euclidean geometries (by Lobatschewsky in 1829 and Bolyai in 1832) shattered infallibilist conceit.[21]

Among the charges *against* fallibilism is that "anything goes" and/or that anyone's ideas are as good as anyone else's. Another is that fallibilists believe that social forces mold mathematics, so it is shaped by the fad of the day rather than by its own logical progression.

Paul Ernest, a specialist in mathematics education at the University of Sussex in England and the editor of a mathematics education journal, maintains that these claims and conclusions are caricatures, that no fallibilist of any worth would subscribe to them. "Fallibilism," he writes, "does not mean that some or all of mathematics may be false (although Gödel's incompleteness results mean that we cannot eliminate the possibility that mathematics may generate a contradiction). . . . A second criticism leveled at fallibilism is that if mathematics is not absolutely necessary then it must be arbitrary or whimsical."

He adds:

Just as realists often caricature the relativist views of social constructivists in science,[22] so too the strengths of the fallibilist views
are not given enough credit. For although fallibilists believe that
mathematics has a contingent, fallible and historically shifting
character, they also argue that mathematical knowledge is to a
large extent necessary, stable and autonomous. Once humans
have invented something by laying down the rules for its existence, like chess, the theory of numbers, or the Mandelbrot set,
the implications and patterns that emerge from the underlying
constellation of rules may continue to surprise us. But this does
not change the fact that we invented the "game" in the first place.
It just shows what a rich invention it was. As the great eighteenth
century philosopher Giambattista Vico said, the only truths we
can know for certain are those we invented ourselves. Mathematics is surely the greatest of such inventions.[23]

Mathematics Is Both Invention *and* Discovery

As we might expect, there are also some people who tread a middle
ground, who believe that mathematics is both discovered and
invented. Among those in the "both" column are Henri Poincaré and
Charles Hermite, who had been Poincaré's teacher. Poincaré, for
example–whose essay "Mathematical Creation" seemed to support
the idea of fallibilism–also wrote an essay titled "Mathematical
Discovery."[24] Hermite's point of view was curious. Probably due to
his religious convictions, he was annoyed with Cantor for creating
certain objects rather than merely discovering them when God saw
fit for this to occur. In other words, Cantor was trying to penetrate
areas, as in his work with the infinite, that God alone should deal
with and would reveal in his own good time.

In 1902, Bertrand Russell wrote, "Not only is mathematics
independent of us and our thoughts, but in another sense we and
the whole universe of existing things are independent of mathematics." Yet on the very next page of the same essay ("The Study of

Mathematics"), he also wrote, "Reason cannot dictate to the world of facts, but the facts cannot restrict reason's privilege of dealing with whatever objects its love of beauty may cause to seem worthy of consideration. Here, as elsewhere, we build up our own ideals out of the fragments to be found in the world; and in the end it is hard to say whether the result is a creation or a discovery."[25] So we'll have to put Mr. Russell in the undecided column.

Furthermore, even Barrow, an avowed Platonist, is aware that the answer to Kline's question is neither simple nor obvious. He asks, for example, "Where is this other world and how do we make contact with it? How is it possible for our mind to have an interaction with the Platonic realm so that our brain state is altered by that experience? Many mathematicians of the Platonic persuasion are strongly influenced by the fact of their own and others' intuition. They have experience of just 'seeing' that certain mathematical theorems are true which makes it feel that they have suddenly come upon mathematical truth by a faculty of 'intuition' that is tantamount to discovery."[26]

This factor of intuition, oddly, is one of the foundation stones of absolutist thinking—that is, that mathematical truths are discovered via the intuition of the mathematician, and that these are then established as correct by various methods of proof.

Barrow continues, "This non-sensory awareness of abstract mathematical structures is a faculty that varies widely, even amongst mathematicians, and so the Platonist must regard the best mathematicians as possessing a means of making contact with the Platonic world more often and more clearly than other individuals."[27]

Roger Penrose, a professor of mathematics at the University of Oxford in England, says in his book *The Emperor's New Mind* (1989) that he is an adherent of the discovery idea in mathematics, but he adds a twist. Perhaps, he says, the matter is not so straightforward, and he suggests that

> There are things in mathematics for which the term "discovery" is indeed much more appropriate than "invention." . . . These are the cases where much more comes out of the structure than is put into it in the first place [for example, Mandelbrot structures]. One may take the view that in such cases the mathematicians have

stumbled upon "works of God" [à la Cantor]. However, there are other cases where the mathematical structure does not have such a compelling uniqueness, such as when, in the midst of a proof of some result, the mathematician finds the need to introduce some contrived and far from unique construction in order to achieve some very specific end. In such cases no more is likely to come out of the construction than was put into it in the first place, and the word "invention" seems more appropriate than "discovery." These are indeed just "works of man." On this view, the true mathematical discoveries would, in a general way, be regarded as greater achievements or aspirations than would the "mere" inventions.[28]

Is There a Crisis in Education?

The battle between the absolutists and the fallibilists goes on in areas other than the high-toned philosophy of mathematics, perhaps most furiously in the world of mathematics education.

Although a variety of studies seem to show that absolutism is still the majority point of view, the fallibilists have worked at least some of their positions into the mathematics curriculum, particularly in England and the United States. One of the important changes, as I stated earlier, was that set-theoretic notation and axiomatics were incorporated into the teaching of high school mathematics in the 1960s.

H. Wu, a mathematics professor at the University of California/ Berkeley, writes:

By the time the idea of the latest reform took hold in 1986—the year NCTM [National Council of Teachers of Mathematics] convened its first meeting to draft the NCTM Standards—the concept of a "proof" in the traditional curriculum had either become non-existent or degenerated into meaningless ritual. For those who went to school in the 40s and 50s, such a statement may come as a surprise, as not a few of us had been charmed by Euclidean geometry—the essence of proofs—into becoming mathematicians.

Yet Euclidean geometry is now perhaps the most vilified portion of school mathematics. What happened? The mathematics curriculum in the schools went through the New Mathematics of the 60s and the Back-to-Basics Movement of the 70s and emerged oversimplified and dumbed down.[29]

Thus in the United States and England, particularly, a loud cry has been raised that our students' capability in mathematics has been diminishing over the last couple of decades. This has happened at the same time that major changes have taken place in the mathematics teaching world. This does not prove that the pedagogical changes are to blame, but the suspicion is strong among many of the remaining absolutists.

Such a challenge to fallibilistic thinking could not remain without counter challenge. Paul Ernest, for example, says, "Their complaints [that is, the absolutists'] are old hat. School mathematics has been criticized since the Great Exhibition of 1851. Boer War defeats were blamed on poor mathematics results. This century, our relative industrial decline has also been blamed on school maths and science."

The real problem, Ernest argues, is that "Too much A level and university maths teaching is outdated and boring." The problem, it would seem, is not that there has been too much change, but that there has not been enough.

Students are expected to learn facts and skills and to regurgitate them for exams, rather than experiencing the excitement of doing and applying real maths. To claim that pupils are deprived of the opportunity to experience "real mathematics" and are being offered a watered down substitute instead is just nonsense.

The problem, on the contrary, is that they are expected to do meaningless and repetitious exercises. . . . Its sorry state is not because it has become too "soft" but because it has failed to generate pupils' enthusiasm and interest. Why are exciting new ideas like Fractals and Chaos Theory not included?

He adds later on,

The unquestioned absolutist view has a lot to do with putting girls off.

If there is a crisis in mathematics I would argue it lies with the attitude of those university mathematicians who locate the problem anywhere except with themselves. Neither back-to-basics nor soft-centred progressivism will solve the problems that maths teaching faces. Teacher educators and researchers like myself acknowledge that we all need to do better and to find out more.[30]

In fact, not everyone even agrees that there is a decline. In some quarters, there are basic disagreements about what pupils should learn in mathematics; what else they should and could be learning at the same time; and what skills they need to learn while in school and what they could gather as they need it later on.

Alan H. Schoenfeld, a professor of education at the University of California/Berkeley, feels that it is still early in the game and that we have only preliminary data. Nevertheless, he writes, "Those data indicate that the first few steps of reform seem to be going in the right direction." In basic agreement with Ernest, he argues that we need to do considerably more, in several areas, including not only curriculum, but in the teaching community and in development of improved methods of assessment.[31]

Pedagogy and Philosophy

The furious debate continues, and it brings us back to our original question. For there is a very close link between the teaching/learning of mathematics and its philosophy/epistemology. Thus Reuben Hersh, writing in *Advances in Mathematics*, maintains, "One's conception of what mathematics *is* affects one's conception of how it should be presented. . . . The issue, then, is not, What is the best way to teach? but, What is mathematics really all about?"[32]

Birgit Pepin, of the Open University in the United Kingdom, looked at mathematics teaching in England, France, and Germany. She argues that "Many of the conditions that exert influence on

without anyone's noticing

human thought and practice within classrooms are neither visible nor readily identifiable. Rather, these forces are the unseen, sometimes 'unperceived,' and often unvoiced principles, philosophies and beliefs that ~~unwittingly~~ penetrate the educational enterprise."[33]

Again, developments in science have provided a model. René Thom points out that science education has been drawing its inspiration from developments in the philosophy of science for a number of years. He adds, "Educational researchers [in mathematics] are becoming increasingly aware of the epistemological foundations of their methodologies and inquiries. . . . In fact," he concludes, "whether one wishes it or not, all mathematical pedagogy, even if scarcely coherent, rests on a philosophy of mathematics."[34]

An interesting classroom experiment was reported in 1992 by Geoffrey Roulet at Queen's University in Kingston, Ontario. A group of student teachers wrote position papers in which they stated their positions on aspects of the subject of mathematics, the student, the teacher, and the relationship of mathematics learning to society. The majority took what amounted to an absolutist position, which argued that the subject should be made relevant to the students' needs, and that this should be done by solving practical problems using concepts that they have been taught and have practiced. A smaller group took a more fallibilist view, one that had more to do with "self-discovery." In general, however, Roulet states quite clearly that "the students' images (philosophies) of mathematics clearly generated corresponding views of teaching." Only then does Roulet's fallibilist leaning become clear: "Moreover the dominant, toolkit view leads to teaching practice that is unlikely to capture pupils' imagination and set them on a path of problem posing and solving."[35]

Finally, Paul Ernest argues, "The philosophy of mathematics education is not about developing a curriculum, but about a theoretical foundation on which curricula might be developed."[36]

And so we are back to our original question—is mathematics discovered or invented? Or, alternatively, are the absolutists or the fallibilists right? We may find that a solid answer, if there is one, will be slow in coming, for the result will both influence and will depend upon the world of mathematics teaching. After all, the students of today will be the curriculum planners of tomorrow.

Epilogue

Years ago, I had lunch with the noted science fiction author Arthur C. Clarke. It was not long after the appearance of his wonderful movie *2001: A Space Odyssey*. I had been wondering about the basic idea of the movie, and I talked with him about it. For example, in an early scene showing prehistoric humans, a puzzling black monolith appears. Toward the end of the movie, in another scene with far more advanced humans, again we see this puzzling black monolith.

I asked Clarke, "Were you suggesting a kind of circularity?"

He thought for a moment, then said, "Yes and no. Yes, there is an idea of circularity, but I didn't want to suggest that we always come back exactly to where we were before. The result is more of a spiral than a circle, with each cycle showing some sort of advance over its predecessor."

The idea has stayed with me ever since, and I think it can be applied nicely to the course that mathematics has taken over the years of its development.

From the earliest days, important revisions of mathematical principles have almost always come after periods of uncertainty, often when contradictions appeared and had to be dealt with. The discovery of irrationals in Pythagorean times led to a crisis in theorems involving proportions. Eudoxus of Cnidus (early 4th century B.C.) got his fellow mathematicians through that crisis with his theory of proportions (equal ratios).

These steps haven't been taken in nice, evenly divided periods. The ups and downs have been uneven, but ups and downs there have been, even within the relatively short couple of hundred years covered in this book. For example, many mathematicians were unhappy with Leibniz's use of infinitesimals for his calculus. It would take almost two centuries before a new formulation, nonstandard analysis, would deal with the problem and permit the calculus to rest comfortably.

More recently, we have had Cantor's introduction of set theory and his work with infinity, Russell's paradox, and Gödel's incompleteness theorem. Each has led to much discomfort and much useful mathematics in response.

Morris Kline, that wily commentator, argues, "The only mathematicians who could retain some composure and smugness from 1931 on, during which time the results [of Gödel's theorem] were breaking the hearts of the logicists, formalists, and set-theorists, were the intuitionists. All the play with logical symbols and principles which taxed the minds of intellectual giants was to them nonsense."[1]

Interestingly, in the same year that Russell came up with his paradox, he also wrote, "One of the chiefest triumphs of modern mathematics consists in having discovered what mathematics really is."

Kline wrote in 1980:

These words strike us as naive today. Beyond the differences in what is accepted as mathematics today by the several schools, one may expect more in the future. The existing schools have been concerned with justifying the existing mathematics. But if one looks at the mathematics of the Greeks, of the 17th century, and of the 19th century one sees dramatic and drastic changes. The several modern schools seek to justify the mathematics of 1900. Can they possibly serve for the mathematics of the year 2000?

The intuitionists do think of mathematics as growing and developing. But would their "intuitions" ever generate or give forth what had not been historically developed? Certainly this was not true even in 1930. Hence it seems that revisions of the foundations will always be needed.[2]

Ernst Snapper, a professor emeritus at Dartmouth, says it is even worse than that. He claims that all three of the schools can be described as undergoing crises in mathematics. For, he argues, none of them has been successful in providing us with a firm foundation for mathematics.[3]

Thus the search for a solid foundation for mathematics is still on; the search for the best approach to the teaching of mathematics is still in progress. With respect to the foundations question, Kline says, "History supports the view that there is no fixed, objective, unique body of mathematics. Moreover, if history is any guide, there will be new additions to mathematics that will call for new foundations."[4]

With each foundational crisis at least, we appear to come back to the same problem. Yet after each crisis, we can expect, or at least hope, that the world of mathematics will have learned something from what transpired before and will emerge both stronger and wiser—after another turn of the spiral.

Notes

Introduction

1. Russell, Bertrand, *Mysticism and Logic* (New York: W. W. Norton, 1929) (essay written in 1902), p. 57.
2. Kline, 1953, p. 105.
3. Hersh, Reuben, *What Is Mathematics, Really?* (New York: Oxford University Press, 1997), pp. 35–37.

Chapter 1. Tartaglia versus Cardano

1. Kline, 1953, p. 109.
2. Also spelled Fiore, and even Florido in some texts.
3. Morley, 1854, vol. 1, p. 218.
4. Ibid., p. 220.
5. Cardan, 1962, p. 142.
6. Ore, 1953, p. 129.
7. Cardan, 1962, p. 15.
8. Ibid., p. 170.
9. Feldman, 1961, p. 162.
10. Ore, 1953, p. 72.
11. Ibid., p. 74.
12. Morley, 1854, vol. 2, p. 233.
13. Ibid., p. 246.
14. Ore, 1953, p. 78.
15. Dunham, 1990, p. 142.
16. Ore, 1953, pp. 84, 85.
17. Cardano, 1993, p. x. Ore, 1953, p. 86.
18. Morley, 1854, vol. 2, p. 249.
19. Wykes, 1969, p. 115.
20. Ore, 1953, p. 68.
21. Cardano, 1993, p. 8.
22. Wykes, 1969, pp. 115–16.
23. Ore, 1953, p. 81.
24. Cardano, 1993, p. xi.
25. Ore, 1953, p. 53.
26. Wykes, 1969, p. 118.
27. Ibid., p. 150.
28. Ibid., p. 173.
29. Wykes, 1969, pp. 174–75.

30. Ore, 1953, pp. 49–50.
31. In Ore, 1953, p. 52.
32. Ore, 1953, p. 106.
33. Ibid.

Chapter 2. Descartes versus Fermat

1. Descartes, 1954, preface.
2. Mahoney, 1994, p. 171.
3. Descartes, in Midonick, 1965, p. 291.
4. Ibid., p. 292.
5. Gaukroger, 1995, p. 100. From Descartes' *Rules for the Direction of the Mind* (1619–1628), translated by John Cottingham and Gaukroger.
6. Descartes, in Hutchins, ed., 1952, p. 47.
7. Gaukroger, 1995, p. 309.
8. Descartes in Midonick, 1965, p. 292.
9. Neither Descartes nor Fermat actually used the sine function. It is used by modern writers because of its ease of use and clarity of description. Seventeenth-century workers referred to lines in a geometrical construction that represent the desired answers and that are equivalent to the sine function.
10. Earlier, in the 1620s, Willibrord Snell, a Dutch professor of mathematics, had developed a similar law, but he had done so on the basis of his own experiments.
11. Descartes, 1954, p. 22. (Orig. pub. 1637.)
12. Coolidge, 1940, p. 122. A locus can be defined as a set of points that satisfies a given condition. For example, the locus of a set of points on a plane equidistant from a point on that plane is a circle. A conic may be defined as the plane curve formed by the intersection of a plane with a right circular cone. The conics are generally taken to be the circle, the ellipse, the parabola, and the hyperbola.
13. Descartes, 1954 (1637), p. 21.
14. Ibid., p. 2.
15. Ibid., p. 6.
16. Grosholz, in Gaukroger, 1980, p. 159.
17. Descartes, 1954 (1637), p. 10, footnote 18. This refers to Descartes' own solution to the Pappus problem, which has been translated into English. Other treatments include Gaukroger, 1995, pp. 210–17; Mahoney, in Gillispie, 1971, pp. 57–58; Grosholz, in Gaukroger, 1980, pp. 157–59; Katz, 1993, pp. 399–404; and Hollingdale, 1989, pp. 130–36. See Boyer, 1991, pp. 336–46, for a more general treatment, and Scott, 1976, pp. 84–133, for a more extended treatment.
18. Fermat, *Oeuvres*, vol. 2, p. 110. Translated by Daniel Curtin.
19. Mahoney, 1994, p. 173.
20. Ibid., pp. 389, 390.
21. Shea, 1991, pp. 292–93, footnote 37.
22. Descartes to Constantin Huygens, August 19, 1638.
23. Descartes to Mersenne, around June 29, 1638.
24. Descartes to Mersenne, July 27, 1638, and April 30, 1639.
25. Descartes to Mersenne, March 4, 1641.
26. Descartes to Mersenne, around September 1641.
27. Descartes to Mersenne, December 1638. Fermat *Oeuvres*, vol. 4, p. 109. With help from James Nicholson and Marshall Hurwitz.
28. Scott, 1976, p. 87.

29. Descartes to Mersenne dated March 1, 1638. Translated by Eric Simon.
30. Ibid., p. 170.
31. See, e.g., Aczel, 1996, and Singh, 1997.
32. Fermat to Mersenne, December 1637; Fermat *Oeuvres,* vol. 2, p. 116. Translated by Daniel Curtin.
33. Fermat to Mersenne, February 1638. Fermat *Oeuvres*, vol. 2, pp. 132, 133. Translated by Daniel Curtin.
34. Fermat to Mersenne, April 20, 1638. Fermat *Oeuvres*, vol. 2, pp. 136, 137. Translated by Daniel Curtin.
35. Descartes to Mersenne, June 29, 1638; in Mahoney, 1994, p. 192.
36. Translation in Mahoney, 1994, p. 58.
37. Bell, 1937, p. 64.
38. Ball, 1960, p. 293.
39. Mahoney, 1994, p. 23.
40. Barner, 2001, p. 15.
41. Barner, 2001, pp. 15–16.
42. See, e.g., Scott, 1976, p. 85; Mahoney, 1994, pp. 27–28, 181; and Coolidge, 1940, p. 119. Gaukroger points out that Descartes' early notation was clumsy and derived from that of Clavius, and that this "indicates that he had not read Viète before developing his own account." (Personal communication, January 15, 2004.)
43. Mahoney, 1994, p. 181.
44. Ibid., p. 388.
45. Ibid.
46. From translations by both J. D. Nicholson, an independent scholar, and Daniel J. Curtin, a professor of mathematics at Northern Kentucky University.
47. Fermat to Clerselier, March 10, 1658. Translated and elucidation by James Nicholson.
48. *Horace, the Odes and Epodes*, with an English translation by Charles E. Bennett (New York: The Macmillan Company, 1914). Bennett was a famous translator of Latin works.
49. Sabra, 1981, p. 130.
50. v_i = velocity of the incident light, and v_r = velocity of the refracted light. Sabra, 1981, p. 149.
51. Sabra, 1981, p. 135.
52. *Ad Locus Planos et Solidos Isagoge* (Introduction to Loci Consisting of Straight Lines and Curves of the Second Degree).
53. Letter to Marin Cureau de la Chambre, January 1, 1662. Translated by Eric Simon.

Chapter 3. Newton versus Leibniz

1. Boorstin, 1991, p. 413.
2. *Philosophical Review*, vol. 52, 1943, p. 366.
3. It's worth noting that even much later on, when his approach to the differential calculus, which he called fluxions, was finally published in 1736 and 1742, it was still titled *Methodus fluxionum et serierum infinitorum*.
4. Hollingdale, 1989, p. 256.
5. More, 1962, p. 394.
6. More, 1962, p. 575.
7. Hall, 1980, p. 140.
8. More, 1962, p. 582.

9. This can be translated as a "Flying Sheet," commonly issued by politicians without date or place of publication.

10. There is some confusion with dates here. The date printed on the article pages is 1714. In England until 1756, however, the year officially ended on March 25. So their February 1714 would be February 1715 to us and would be stated as such in some historical sources.

11. Anonymous (Newton), 1714, p. 139.

12. Hall, 1980, pp. 39, 187; Youschkevitch, 1974, p. 47.

13. See, e.g., More, 1962, pp. 592–94, for further details on this.

14. Merz, 1884, p. 126.

Chapter 4. Bernoulli versus Bernoulli

1. Publisher's Web site, page on Jakob: www.springeronline.com/sgw/cda/frontpage/0,11855,4-40295-2-122580-0,00.html

2. Thanks here to Prof. Siegmund Probst, at the University of Hannover, who helped me sort out the catenary puzzle.

3. Kline, 1972, p. 473. Original citation: Johann Bernoulli, *Der Briefwechsel von Johann Bernoulli* (Basel, Switzerland: Birkhäuser Verlag, 1955), pp. 97–98.

4. E.g., for the catenary: Kline, 1972, pp. 472–73; for the velaria: Cajori, 1980, p. 221, and Fellman and Fleckenstein, 1970, p. 53.

5. Ball, 1960, p. 366.

6. Bell, 1937, p. 134; *Encyclopædia Britannica*, vol. 2, 1998, p. 154; Durant, 1963, p. 501; Calinger, 1995, p. 419.

7. Cajori, 1980, p. 221.

8. Thiele, 1997, p. 261.

9. Personal communication, October 13, 2004.

10. *Encyclopædia Britannica*, vol. 2, 1998, p. 154.

11. Thiele, 1997, p. 263.

12. Ibid., p. 266. Translated by Fred Stern. Original citation: Bernoulli, Joh: Briefwechsel. Bd. 1. S. 435f. (Brief vom 28. 5./7. 6. 1701).

13. Ibid.

14. Hofmann, 1970, p. 50.

15. Struik, 1969, p. 393.

16. Ibid., p. 398.

17. Historians' use of the brothers' first names seems to be a free-for-all. Jakob is variously referred to as Jacob, James, and Jacques, while Johann may also be John or Jean.

18. Bell, 1945, p. 377.

19. Kline, 1972, p. 575.

20. Smith, 1959, p. 644.

21. Hollingdale, 1989, p. 288.

22. O'Connor and Robertson, 1998 ("Johann"), p. 4.

23. Smith, 1959, p. 645.

24. Hollingdale, 1989, p. 288.

25. Smith, 1959, p. 645.

26. Hollingdale, 1989, p. 289.

27. Ibid.

28. Bernoulli, in Smith, (vol. 2), 1959, pp. 649, 654.

29. Fellman and Fleckenstein, 1970, p. 53.

30. More details in Fellman and Fleckenstein, 1970, p. 53.
31. Struik, 1969, p. 320.
32. Sierksma, 1992, p. 26.
33. Ibid.
34. Ibid., pp. 25, 27.
35. Ibid., p. 28.
36. Rusting, 1990, p. 34.
37. Newman, 1956, vol. 2, p. 772.
38. See, e.g., Straub, 1970, p. 37.
39. See, e.g., Wikipedia Encyclopedia online, p. 1, and O'Connor and Robertson, 1998 ("Johann"), p. 4.
40. Thiele, 1997, p. 267. Original citation: Fuss, P. H.: *Correspondance mathématique et physique.* Tome 2. St.-Pétersburg, 1843. S. 530.
41. Galton, 1952, p. 195.

Chapter 5. Sylvester versus Huxley

1. Desmond, 1997, p. 377.
2. Ibid., p. xvii.
3. Huxley, 1869, p. 126.
4. In Huxley, 1900, p. 392.
5. Huxley, op. cit., 1869, p. 126.
6. Huxley, 1870, p. 146.
7. See Huxley, 1869, in the bibliography.
8. Macfarlane, 1916, p. 100.
9. Desmond, 1997, p. 339.
10. Bibby, 1972, p. 1.
11. Huxley, 1854, p. 63.
12. Ibid., pp. 57–58.
13. Huxley, 1856, p. 307.
14. Paradis, 1978, pp. 57–58.
15. Huxley, 1868, pp. 130–65.
16. Huxley, 1869, pp. 125–26, 130–31.
17. In Huxley, 1900, p. 189.
18. See, e.g., Hellman, 1998, pp. 84–88; Desmond, 1997, pp. 276–81.
19. Obtained online. See http://aleph0.clarku.edu/huxley/guide5.html.
20. Ashforth, 1969, p. 54.
21. Huxley, 1869, p. 333.
22. See, e.g., Hellman, 1999, pp. 110–17, or Ashforth, 1969, pp. 53–54.
23 Huxley, 1870, p. 130.
24. Original quote in French. Translated by Eric J. Simon.
25. Huxley, op. cit., p. 146.
26. Bell, 1937, p. 385.
27. Parshall, 1998, p. 1.
28. Bell, 1937, p. 387.
29. *Encyclopedia Americana*, vol. 26, p. 158.
30. In Parshall, 1998, p. 80.
31. Ibid.
32. Bell, 1937, p. 385.
33. Newman, 1956, p. 340.

34. Ibid.
35. Sylvester, 1908, p. 651. (Orig. pub. 1869.) His talk is also reproduced in Midonick, 1965, pp. 759–68.
36. Ibid., p. 652.
37. Ibid., p. 653.
38. Ibid.
39. Ibid., p. 654.
40. Ibid.
41. Ibid.
42. Ibid., pp. 654–55.
43. Ibid., pp. 655–56.
44. Ibid., pp. 656–57.
45. Ibid., p. 657.
46. Ibid.
47. Ibid., pp. 657–58.
48. Desmond, 1997, p. xiii.
49. Bibby, 1972, p. 122.
50. Ibid., p. 113.
51. Macfarlane, 1916, p. 113.
52. Huxley, 1876, p. 240.
53. Huxley, 1874, p. 207.
54. Huxley, 1882, p. 177.
55. In Bell, 1937, p. 396.

Chapter 6. Kronecker versus Cantor

1. Galileo Galilei, *Dialogues Concerning Two New Sciences* (New York: McGraw-Hill, 1963), p. 29. (Orig. pub. 1638.)
2. Kline, 1972, p. 993.
3. Hollingdale, 1989, p. 359.
4. Dauben, 1990, p. 4.
5. Ibid., p. 1.
6. Cajori, 1980, p. 362. (Orig. pub. 1893.)
7. Barrow, 1992, p. 200.
8. Aczel, 2000, p. 77.
9. Bell, 1937, p. 562.
10. Bell, 1937, p. 559.
11. Dauben, 1990, p. 147.
12. Galileo. *Dialogues Concerning Two New Sciences*, p. 31.
13. Good descriptions of the proof can be found in Dauben, 1990, pp. 50–54; Aczel, 2000, pp. 114–16; and Dunham, 1990, pp. 259–61.
14. Meschkowski, 1971, p. 54.
15. Wallace, 2003, p. 259.
16. The term *dangerous world of mathematical insanity* is Eric Temple Bell's colorful expression (Bell, 1937, p. 570). As we'll see, the term can have a double meaning. Joseph Dauben feels that it plays unfairly on Cantor's later battle with manic depression (personal communication, January 4, 2006). Kronecker himself certainly used such terms as *humbug*, however.
17. Kline, 1972, p. 995.
18. Collins and Restivo, 1983, p. 217.

19. Dauben, 1990, p. 134.
20. Ibid., p. 135.
21. Kline, 1972, p. 999.
22. Bell, 1937, p. 560.
23. Grattan-Guinness, 1971a, p. 382.
24. Dauben, 1990, p. 136.
25. Ibid.
26. Ibid., p. 313.
27. For anyone who wishes to pursue this aspect of Cantor's life further, and can read German, a book of his correspondence with sympathetic theologians has recently been published: *Kardinalität und Kardinäle: wissenschaftshistorische Aufarbeitung der Korrespondenz zwischen Georg Cantor und katholischen Theologen seiner Zeit*, ed. Christian Tapp (Stuttgart: Steiner, 2005).
28. In Aczel, 2000, p. 162.
29. Sweeney, 2001, p. 565.
30. Dunham, 1990, p. 283.
31. Bell, 1937, p. 570.
32. Ibid.
33. Edwards, 1987, p. 31.
34. Ibid., p. 34.
35. Ibid., pp. 33–34.
36. Barrow, 1992, p. 201.
37. Kline, 1972, p. 1198.
38. Edwards, 1987, pp. 34–35.
39. Personal communication, September 14, 2003.
40. Collins and Restivo, 1983, p. 218.

Chapter 7. Borel versus Zermelo

1. Moore, 1982, p. 42.
2. Maddy, 1990, p. 117.
3. Bell, 1945, p. 484.
4. See, e.g., Moore, 1982, pp. 93–141.
5. Jervell, 1996, p. 96.
6. Russell, in Heinzmann, 1986, pp. 72–73.
7. Moore, 1982, p. 313.
8. Jervell, 1996, p. 96.
9. Moore, 1982, p. 159.
10. Kline, 1980, p. 211.
11. Maddy, 1990, p. 118.
12. Ibid.
13. Ibid., p. 121.
14. Moore, 1982, p. 178.
15. Ibid.
16. Ibid.
17. Ibid., p. 167.
18. Russell, in Whitehead, 1997, p. viii.
19. Dauben, 1990, p. 267.
20. Kline, 1980, pp. 211–12.
21. Dauben, 1990, p. 268.

Chapter 8. Poincaré versus Russell

1. Russell, 1959, pp. 76–77.
2. Ibid., pp. 75–76.
3. Ibid., p. 74.
4. Russell, *Mysticism and Logic,* 1918(?), pp. 70–71.
5. Russell, 1938, p. v.
6. Nordmann, in Jones, 1966, p. 619.
7. Russell, 1967, pp. 17–18.
8. Ibid., pp. 37–38.
9. Moorhead, 1992, p. 42.
10. Russell, 1967, p. 87.
11. Ibid., p. 90.
12. *Grundlagen einer allgemeinen Mannichfaltigkeitslehre (Foundations of a General Theory of Sets),* 1883.
13. Roughly, *Concept Notation,* Frege's first publication on logic (1879).
14. Russell, 1967, p. 91.
15. Ibid., p. 187.
16. Kline, 1972, p. 1192.
17. Russell, 1938, p. xvi.
18. O'Connor and Robertson, 2002, online.
19. See, e.g., Kitcher and Aspray, "An Opinionated Introduction," in Aspray and Kitcher, 1988, pp. 14–16.
20. Russell, 1967, pp. 217–18.
21. Russell, 1919, p. 5.
22. Burton, 1991, p. 655.
23. Russell, 1973, p. 145.
24. Newman, 1956, p. 1377.
25. Kline, 1972, p. 1003.
26. O'Connor and Robertson, 2003, p. 5.
27. Quoted by Nordmann, in Jones, 1966, p. 619.
28. Kline, 1980, p. 233.
29. Ibid.
30. Poincaré, 1906, in Poincaré, 1946, p. 435.
31. Ibid., pp. 449, 451, 452.
32. Ibid., pp. 479–80.
33. Russell, 1906, in Russell, 1973, p. 164.
34. Poincaré, 1906, in Poincaré, 1946, p. 480. This text comes from Virgil's *Aeneid* (I.118). It describes the shipwreck of Aeneas's vessel in a storm off the coast of Africa (near Carthage), caused by Neptune.
35. Ibid., p. 472.
36. Russell, 1906, in Russell, 1973, p. 191.
37. Ibid.
38. Ibid., p. 192.
39. Ibid., pp. 213–14.
40. Poincaré, 1906, in Poincaré, 1946, pp. 484, 485.
41. See Russell, 1908.
42. See editor's summary in Russell, 1973, p. 133.
43. Russell had earlier used predicative for properties that define classes. In 1906 Poincaré proposed that properties are predicative only if they contain no vicious circles.

Later, both Poincaré and Russell defined predicative to mean containing no vicious circles.

44. Goldfarb, 1988, p. 79.
45. Russell, 1910, in Russell, 1973, p. 215.
46. Ibid., pp. 244–50.
47. Ibid., p. 252.
48. Russell, 1938, p. v.
49. Ibid., p. xii.
50. Ibid., pp. xii–xiii.
51. Ibid., p. xiv.
52. Kline, 1980, p. 311.
53. Russell, 1938, p. v.
54. Detlefsen, 1993, pp. 49 and 28.
55. See, e.g., Van Evra, 2003, p. 387.
56. See, e.g., Lambek, 1994, or Linsky and Zalta, 2004.
57. See, e.g., Lambek, 1994, p. 59; Broadbent, 1975, p. 15; Simonis, 1999, pp. 172–73; Kline, 1972, p. 1197.

Chapter 9. Hilbert versus Brouwer

1. Russell, 1919 (1993), quoted in Slater, 1994, introduction.
2. Peckhaus, 2003, p. 3.
3. This was actually number 6 in his *published* list of 23 problems.
4. In Barrow, 1992, p. 112.
5. In Burton, 1991, p. 657.
6. Weyl, in Reid, 1970, p. 264.
7. See, e.g., chapter 8 and/or Barrow, 1992, p. 114n.
8. In Barrow, 1992, pp. 195–96.
9. Dirk van Dalen, personal communications, August 11 and 16, 2005.
10. Van Dalen, 1990, p. 18.
11. Reid, 1970, p. 6.
12. Ibid.
13. Reid, 1970, p. 94.
14. In Simpson, 1986, p. 3. From Hilbert, "On the Infinite," pp. 367–92 in van Heijenoort (chapter 8).
15. Simpson, 1986, p. 4.
16. Van Dalen, 1999, pp. ix–x.
17. He also says that he has been defending this idea since 1907. Paper presented to the Royal Academy of Sciences, December 18, 1920. In Mancosu, 1998, p. 23.
18. Reid, 1970, p. 149.
19. In O'Connor, 2003, p. 3.
20. Dirk van Dalen, personal communication, August 16, 2005.
21. Mancosu, 1998, p. 120.
22. Hilbert, "The New Grounding of Mathematics. First Report," in Mancosu, 1998, p. 198.
23. Ibid., pp. 199–201.
24. Ibid., pp. 202–3.
25. Dirk van Dalen, personal communication, September 21, 2005.
26. Van Dalen, 2005, p. 640.

27. Paraphrased by Nyikos, 2004, p. 2.

28. Brouwer, "Intuitionist Reflections on Formalism" (1928), in Mancosu, 1998, p. 42.

29. In Mancosu, 1998, p. 10.

30. Van Dalen, 2005, p. 601. Many of the following quotes can also be found in van Dalen, 1990.

31. Ibid., p. 613.

32. Ibid., p. 603.

33. Ibid., p. 605.

34. Ibid., p. 614.

35. Which he later did. Working with a Dutch publisher, he founded the *Compositio Mathematica* in 1934.

36. Van Dalen, 2005, p. 613.

37. Personal communication, August 16, 2005.

38. This is the title of a Greek play of unknown authorship. A German medieval version was penned by Rollenhagen.

39. Van Dalen, 2005, p. 619.

40. Dirk van Dalen, personal communication, September 16, 2005.

41. Van Dalen, 1990, p. 31.

42. Reid, 1970, p. 187.

43. Intuitionism is sometimes referred to as constructivism. For example, intuitionists do not accept transfinite ordinals beyond the second number class. That is to say, one must be able to count ordinals more or less like the ordinary integers. Nor can intuitionists accept Cantor's higher cardinals, because these cannot be constructively arrived at in the intuitionist's sense of mental creation. This is in contrast to the Cantorian practice, where anything that can be formulated consistently exists.

44. Snapper, in Swetz, 1994, p. 707.

45. Ibid.

Chapter 10. Absolutists/Platonists versus Fallibilists/Constructivists

1. Ernest, 2004 (in section "Linking Philosophies of Mathematics and Mathematical Practice").

2. Thom, 1990. See also Thomas S. Kuhn, *The Structure of Scientific Revolutions*, 2nd ed. (Chicago: University of Chicago Press, 1970). (Orig. pub. 1962.)

3. Ibid.

4. Kline, 1980, p. 323.

5. Barrow, 1992, p. 257.

6. In Kline, 1980, p. 322.

7. Ibid.

8. Hardy, 1967, p. 123. (Orig. pub. 1941.)

9. In Kline, 1980, p. 323.

10. Kline, 1980, p. 323.

11. Barrow, 1992, p. 258.

12. Ibid.

13. Ibid., p. 263.

14. Ibid., p. 260.

15. Ibid., p. 177.

16. Poincaré, in Newman, 1956, pp. 2043, 2045.

17. Personal communication, September 15, 2003.

18. Reid, 1970, p. 175.

19. Bridgman, *The Logic of Modern Physics* (New York: Macmillan, 1927), p. 60.

20. Kasner, Edward, and James Newman, *Mathematics and the Imagination* (New York: Simon and Schuster, 1980), p. 359. (Orig. pub. 1940.)

21. Lakatos, I., *Proofs and Refutations: The Logic of Mathematical Discovery* (New York: Cambridge University Press, 1976), p. 139.

22. For some general arguments against constructivism, see Rowlands and Carson, 2001.

23. Ernest, 1999, pp. 2, 3, and 4.

24. In Rapport and Wright, 1963, pp. 128–37. Original in Poincaré, 1946. (Orig. pub. 1913.)

25. Russell, 1957, pp. 65, 66.

26. Barrow, 1992, pp. 272–73.

27. Ibid., p. 273.

28. Penrose, 1989, pp. 96–97.

29. Wu, 1996, p. 1532.

30. Ernest, 1995, p. T15.

31. Schoenfeld, 2002, p. 14.

32. Hersh, 1979, p. 32.

33. Pepin, 1999, p. 128.

34. Thom, 1990.

35. Roulet, 1992.

36. Ernest, 1994, p. 5.

Epilogue

1. Kline, 1980, p. 276.

2. Ibid., p. 277.

3. In Swetz, 1994, pp. 697, 707.

4. Kline, 1980, p. 320.

Bibliography

General Background

Abbott, David. *The Biographical Dictionary of Scientists: Mathematicians.* New York: Peter Bedrick Books, 1986.

Albers, Donald J., and G. L. Alexanderson, eds. *Mathematical People: Profiles and Interviews.* Cambridge, Mass.: Birkhäuser Boston, 1985.

Aspray, William, and Philip Kitcher, eds. *History and Philosophy of Modern Mathematics.* Minneapolis: University of Minnesota Press, 1988.

Ball, W. W. Rouse. *A Short Account of the History of Mathematics.* New York: Dover, 1960. (Orig. pub. 1908.)

Barrow, John D. *Pi in the Sky: Counting, Thinking, and Being.* New York: Oxford University Press, 1992.

Bell, E. T. *The Development of Mathematics.* 2nd ed. New York: McGraw-Hill, 1945.

———. *Men of Mathematics.* New York: Simon and Schuster, 1937.

Boyer, Carl B. *A History of Mathematics.* 2nd. rev. ed. New York: John Wiley & Sons, 1991.

Bruno, Leonard C. *Math and Mathematicians: The History of Math Discoveries around the World.* Detroit: UXL, 1999.

Burton, David M. *The History of Mathematics: An Introduction.* Boston: McGraw-Hill, 2003. (Orig. pub. 1991.)

Cajori, Florian. *A History of Mathematics.* New York: Chelsea Publishing, 1980. (Orig. pub. 1893.)

Calinger, Ronald. *Classics of Mathematics.* Englewood Cliffs, N.J.: Prentice-Hall, 1995. (Orig. pub. Moore, 1982.)

Collins, Randall, and Sal Restivo. "Robber Barons and Politicians in Mathematics: A Conflict Model of Science." *Canadian Journal of Sociology* 8, no. 2 (Spring 1983): 199–227.

Davis, Philip J., and Reuben Hersh. *The Mathematical Experience.* Boston: Birkhäuser, 1981.

Doxiadis, Apostolos K. *Uncle Petros and Goldbach's Conjecture.* New York: Bloomsbury USA, 2000. (An absorbing novel about a mathematician's compulsion to solve a major mathematical problem.)

Dunham, William. *Journey through Genius: The Great Theorems of Mathematics.* New York: John Wiley & Sons, 1990.

———. *The Mathematical Universe. An Alphabetical Journey through the Great Proofs, Problems, and Personalities.* New York: John Wiley & Sons, 1994.

Durant, Will, and Ariel. *The Age of Louis XIV.* New York: Simon and Schuster, 1963. (Useful background information for the years 1648–1714.)

Fadiman, Clifton. *Fantasia Mathematica*. New York: Simon and Schuster, 1958. (Stories, poetry, and other oddments.)

Galton, Francis. *Hereditary Genius: An Inquiry into Its Laws and Consequences*. New York: Horizon Press, 1952. (Orig. pub. 1869.) (Some historical information about the Bernoulli brothers.)

Gillispie, Charles C., ed. *Dictionary of Scientific Biography*. 16 vols. New York: Scribner, 1970–1980.

Goldstein, Catharine, and Jeremy Gray. "The Roots of Modern Maths (From the Middle Ages to the Enlightenment)," *UNESCO Courier* (November 1989): 42ff.

Grattan-Guinness, Ivor. *The Norton History of the Mathematical Sciences*. New York: W. W. Norton, 1998.

Guedj, Denis. *The Parrot's Theorem*. New York: St. Martin's, 2000. (A combined mystery and quick tour of the history of mathematics.)

Guillen, Michael. *Five Equations That Changed the World: The Power and Poetry of Mathematics*. New York: Hyperion, 1995.

Gullberg, Jan. *Mathematics: From the Birth of Numbers*. New York: W. W. Norton, 1997.

Hellman, Hal. *Great Feuds in Science*. New York: John Wiley & Sons, 1998.

Henderson, Harry. *Modern Mathematicians*. New York: Facts on File, 1996.

Hogben, Lancelot. *Mathematics in the Making*. New York: Crescent Books, 1960.

———. *Mathematics for the Million*. 2nd ed. New York: W. W. Norton, 1940. (Orig. pub. 1937.)

Hollingdale, Stuart. *Makers of Mathematics*. New York: Viking Penguin, 1989.

Hooper, A. *The River Mathematics*. New York: Henry Holt, 1945.

Jones, Bessie Zaban, ed. *The Golden Age of Science: Thirty Portraits of the Giants of 19th-Century Science*. New York: Simon and Schuster, 1966.

Katz, Victor J. *A History of Mathematics: An Introduction*. New York: HarperCollins, 1993.

Kline, Morris. *Mathematical Thought from Ancient to Modern Times*. New York: Oxford University Press, 1972.

———. *Mathematics: The Loss of Certainty*. New York: Oxford University Press, 1980.

———. *Mathematics and the Search for Knowledge*. New York: Oxford University Press, 1985.

———. *Mathematics in Western Culture*. New York: Oxford University Press, 1953.

Laubenbacher, R., and David Pangelley. *Mathematical Expeditions: Chronicles by the Explorers*. New York: Springer-Verlag, 1999.

Macfarlane, Alexander. *Lectures on Ten British Mathematicians of the Nineteenth Century*. New York: John Wiley & Sons, 1916.

Mazur, Barry. *Imagining Numbers: Particularly the Square Root of Minus Fifteen*. New York: Farrar, Straus and Giroux, 2002.

Midonick, H. *The Treasury of Mathematics*. New York: Philosophical Library, 1965.

Muir, Jane. *Of Men and Numbers: The Story of the Great Mathematicians*. New York: Dodd, Mead, 1961.

Nahin, Paul J. *When Least Is Best: How Mathematicians Discovered Many Clever Ways to Make Things as Small (or as Large) as Possible*. Princeton, N.J.: Princeton University Press, 2004.

Newman, James R., ed. *The World of Mathematics. A Small Library of the Literature of Mathematics from A'h-mosé the Scribe to Albert Einstein*. 4 vols. New York: Simon and Schuster, 1956.

Pappas, Theoni. *Mathematical Scandals*. San Carlos, Calif.: Wide World Publishing/Tetra, 1997.

Rapport, Samuel, and Helen Wright. *Mathematics*. New York: Washington Square Press, 1963.

Simonis, Doris, ed. *Scientists, Mathematicians, and Inventors: Lives and Legacies: An Encyclopedia of People Who Changed the World*. Phoenix, Ariz.: Oryx Press, 1999.

Smith, David Eugene. *A Source Book in Mathematics*. 2 vols. New York: Dover, 1959. (Orig. pub. 1929.)

Struik, Dirk J. *A Concise History of Mathematics*. New York: Dover, 1987. (Orig. pub. 1948.)

Struik, D. J., ed. *A Source Book in Mathematics, 1200–1800*. Cambridge, Mass.: Harvard University Press, 1969.

Suzuki, Jeff. *A History of Mathematics*. Upper Saddle River, N.J.: Prentice Hall, 2002.

Swetz, Frank J., ed. *From Five Fingers to Infinity: A Journey through the History of Mathematics*. Chicago: Open Court, 1994.

Tocquet, Robert. *The Magic of Numbers*. New York: A. S. Barnes, 1961.

Turnbull, Herbert Westren. *The Great Mathematicians*. New York: Barnes & Noble, 1993. (Orig. pub. 1961.)

Wiener, Philip P., and Aaron Noland. *Roots of Scientific Thought: A Cultural Perspective*. New York: Basic Books, 1957.

Young, Robyn V., ed. *Notable Mathematicians: From Ancient Times to the Present*. Detroit: Gale, 1998.

Introduction

Crowe, Michael J. "Ten Misconceptions about Mathematics and Its History," in Aspray and Kitcher, 1988, pp. 260–77.

1. Tartaglia versus Cardano

Ashworth, Allan. "Cardano's Solution." *History Today* (January 1999): 46.

Bidwell, James, and Bernard K. Lange. "Girolamo Cardano: A Defense of His Character." *Mathematics Teacher* 64 (January 1971): 25–71.

Cardan, Jerome. *The Book of My Life (De Vita Propria Liber)*. New York: Dover, 1962.

Cardano, G. *Ars Magna or The Rules of Algebra*. New York: Dover, 1993. (Republication of 1968 MIT edition; translation of *Ars Magna,* Nürnberg, 1545, with additions from 1570 and 1663 editions. Translated and edited by T. Richard Witmer.)

Feldman, Richard. "The Cardano-Tartaglia Dispute." *Mathematics Teacher* 54 (March 1961): 160–63.

Field, Judith Veronica. *The Invention of Infinity: Mathematics and Art in the Renaissance*. New York: Oxford University Press, 1997.

Morley, Henry. *The Life of Girolamo Cardano, of Milan, Physician*. London: Chapman and Hall, 1854.

Muir, Jane. "Cardano, 1501–1576." In Muir, 1961, pp. 26–46.

Ore, Øystein. *Cardano, the Gambling Scholar*. With a translation from the Latin of Cardano's book on games of chance, by Sydney Henry Gould. Princeton, N. J.: Princeton University Press, 1953.

Rashed, Roshdi. "Where Geometry and Algebra Intersect." *UNESCO Courier* (November 1989): 36–41.

Waerden, B. L. van der. *A History of Algebra: From al-Khwarizmi to Emmy Noether*. New York: Springer-Verlag, 1985.

Wykes, Alan. *Doctor Cardano, Physician Extraordinary*. London: Muller, 1969.

2. Descartes versus Fermat

Aczel, Amir D. *Fermat's Last Theorem: Unlocking the Secret of an Ancient Mathematical Problem.* New York: Four Walls Eight Windows, 1996.

Barner, Klaus. "Pierre de Fermat (1601?–1665). His Life besides Mathematics." *EMS Newsletter* 42 (December 2001): 12–16. [European Mathematical Society.]

Boyer, Carl. "Analytic Geometry: The Discovery of Fermat and Descartes," *Mathematics Teacher* 37, no. 3 (1944): 99–105.

———. *History of Analytic Geometry.* New York: Scripta Mathematica, 1956.

———. "The Invention of Analytic Geometry." *Scientific American* (January 1949): 40–45.

Coolidge, Julian Lowell. *A History of Geometrical Methods.* Oxford: Clarendon Press, 1940.

Cottingham, John. *Descartes.* New York: Routledge, 1999.

Descartes, René. "Discourse on Method." In Midonick, 1965, pp. 290–96. Translated by John Veitch.

———. "Discourse on the Method of Rightly Conducting the Reason." In Hutchins, 1952, pp. 41–68. Translated by Elizabeth S. Haldane and G. R. T. Ross.

———. *The Geometry of René Descartes with a facsimile of the first edition.* Translated from French and Latin by David Eugene Smith and Marcia L. Latham. New York: Dover, 1954. (Orig. pub. 1925.)

———. *Oeuvres de Descartes*, 2nd ed., 11 vols. Ed. Charles Adam and Paul Tannery. Paris: Paris Librarie Philosophique J. Vrin, 1965–1973.

———. *The Philosophical Writings of Descartes*, trans. John Cottingham et al. 2 vols. New York: Cambridge University Press, 1984.

Gabbey, Alan. "Force and Inertia in the Seventeenth Century: Descartes and Newton." Chapter 10 in Gaukroger, 1980, pp. 230–320 (especially part 2, pp. 243–72, re Descartes' *determinatio*).

Galison, Peter. "Descartes: An Intellectual Biography." *New Republic* (May 13, 1996): 39–46. (Essay review of *Descartes: An Intellectual Biography*, by Stephen Gaukroger.)

Gaukroger, Stephen. *Descartes: An Intellectual Biography.* New York: Oxford University Press, 1995.

Gaukroger, Stephen, ed. *Descartes: Philosophy, Mathematics and Physics.* Totowa, N.J.: Barnes & Noble, 1980. Especially chapter 4, "Descartes' Project for a Mathematical Physics," pp. 97–140.

Grosholz, Emily R. "Descartes' Unification of Algebra and Geometry." Chapter 6 in Gaukroger, 1980, pp. 156–68.

Haldane, Elizabeth S. *Descartes: His Life and Times.* London: J. Murray, 1905.

Hutchins, Robert Maynard, ed. *Bacon, Descartes, Spinoza.* 2nd ed. Vol. 28 of *Great Books of the Western World.* Chicago: University of Chicago Press (Encyclopedia Britannica), 1990.

Indorado, Luigi, and Pietro Nastasi. "The 1740 Resolution of the Fermat-Descartes Controversy." *Historia Mathematica* 16 (1989): 137–48.

Mahoney, Michael S. "The Beginnings of Algebraic Thought in the Seventeenth Century." Chapter 5, in Gaukroger, 1980, pp. 141–55.

———. "Descartes: Mathematics and Physics." In Gillispie, vol. 4, 1971, pp. 55–61.

———. "Fermat, Pierre de." In Gillispie, vol. 4, 1971, pp. 566–76.

———. *The Mathematical Career of Pierre de Fermat, 1601–1665.* Princeton, N.J.: Princeton University Press, 1994. (Orig. pub. 1973.)

Pratter, Frederick. "Descartes: Philosophical Pioneer Biography Brings to Light Intellectual Life in 17th-Century Europe." Review of Gaukroger, 1995. *Christian Science Monitor* (June 15, 1995): 13.

Sabra, A. I. *Theories of Light, from Descartes to Newton.* New York: Cambridge University Press, 1981. (Orig. pub. 1967.)

Scott, J. F. *The Scientific Work of René Descartes.* London: Taylor & Francis, 1976. (Orig. pub. 1952.)

Shea, William R. *The Magic of Numbers and Motion: The Scientific Career of René Descartes.* Canton, Mass.: Science History Publications, 1991.

Singh, Simon. *Fermat's Enigma: The Epic Quest to Solve the World's Greatest Mathematical Problem.* New York: Walker and Company, 1997.

———. "Mathematicians, Romantic Heroes of the Past, Present and Future." Undated, accessed December 13, 2003. www.simonsingh.net/Mathematical_Heroes.html

3. Newton versus Leibniz

Andrade, E. N. da C. "Newton and the Science of His Age." *Nature* 150 (December 19, 1942): 700–706.

———. *Sir Isaac Newton.* London: Collins, 1954.

Anonymous (Newton). "An Account of a Book entitled: *Commercium Epistolicum Collinii et Aliorum, De Analysi promota; published by order of the Royal Society, concerning the Dispute between Mr. Leibnitz and Dr. Keill, about the Right to the Invention of the Method of Fluxions, by some called the Differential Method,* no. 342, p. 173." *Philosophical Transactions* 29 (1714): 116–53.

Berlinski, David. *A Tour of the Calculus.* New York: Pantheon, 1995.

Boorstin, Daniel. *The Discoverers.* New York: H. N. Abrams, 1991.

Boyer, Carl B. *The History of the Calculus and Its Conceptual Development.* New York: Dover, 1959. (Orig. pub. 1949.)

Broad, William J. "Sir Isaac Newton: Mad as a Hatter." *Science* 213 (September 18, 1981): 1341, 1342, 1344. Also letters, November 13, 1981, and March 5, 1982.

Brodetsky, S. "Newton: Scientist and Man." *Nature* 159 (December 19, 1942): 698–99.

Cassirer, Ernst. "Newton and Leibniz." *Philosophical Review* 52, no. 4 (July 1943): 366–91.

Frankfurt, Harry G., ed. *Leibniz: A Collection of Critical Essays.* New York: Doubleday (paperback), 1972. Especially essay on "Leibniz and Newton."

Gleick, James. *Isaac Newton.* New York: Pantheon, 2003.

Grattan-Guinness, Ivor. "The Calculus and Its Consequences, 1660–1750." In Grattan-Guinness, 1998, pp. 234–301.

Guillen, Michael. *Five Equations That Changed the World.* New York: Hyperion, 1995. Especially section on Newton (pp. 9–63).

Hall, A. Rupert, and Laura Tilling, eds. *The Correspondence of Isaac Newton.* Vol. 7, 1718–1727. New York: Cambridge University Press, 1977.

———. *From Galileo to Newton.* New York: Dover, 1981. (Prev. pub. Harper & Row, 1963.)

———. *Philosophers at War: The Quarrel between Newton and Leibniz.* New York: Cambridge University Press, 1980.

Hathaway, Arthur S. "The Discovery of the Calculus" *Science* (July 11, 1919): 41–43.

———. "Further History of the Calculus." *Science* (February 13, 1920): 166–67.

Hofmann, Joseph E. "Leibniz: Mathematics." In Gillispie, vol. 8, 1973, pp. 160–68.

Hogben, Lancelot. "The Newton-Leibniz Calculus." Chapter 10, pp. 212–29, in Hogben, 1960.

Hunt, Frederick Vinton. *Origins in Acoustics: The Science of Sound from Antiquity to the Age of Newton.* New Haven, Conn.: Yale University Press, 1978. (Couple of pages on Newton-Leibniz feud, pp. 146ff.)

Latta, Robert, ed. *Leibniz. The Monadology and Other Philosophical Writings*. London: Oxford University Press, 1898.

Kreiling, Frederick. "Leibniz, Gottfried Wilhelm." In Gillispie, vol. 8, 1973, pp. 149–50.

Manuel, Frank E., "Newton as Autocrat of Science." *Daedalus* (Summer 1968): 969–1001.

Merz, J. T. *Leibniz*. New York: Lippincott, 1884.

Mittelstrass, Jürgen, and Eric J. Aiton. "Leibniz: Physics, Logic, Metaphysics." *Dictionary of Scientific Biography*, vol. 8, 1973, pp. 150–60.

More, Louis T. *Isaac Newton. A Biography*. New York: Dover, 1962. (Orig. pub. 1934.)

Newton, Isaac. *Mathematical Principles of Natural Philosophy*. Chicago: Encyclopedia Britannica, 1955. (Orig. pub. 1687.)

Peursen, C. A. van. *Leibniz*. New York: Dutton, 1970.

Price, Derek J. de Solla. *Little Science, Big Science*. New York: Columbia University Press, 1986. Especially p. 68.

Schrader, Dorothy V. "The Newton-Leibniz Controversy Concerning the Discovery of the Calculus." In Swetz, 1994, pp. 509–22.

Smith, Preserved. *A History of Modern Culture*, 1930, vol. 1 (*The Great Renewal, 1543–1687*), 1930, vol. 2 (*The Enlightenment, 1687–1776*). New York: Henry Holt. (Reprinted 1957 by Peter Smith.)

Spitz, L. W. "Leibniz's Significance for Historiography." *Isis* 13 (1952): 333–48.

Stephenson, Neal. *Quicksilver*. New York: HarperCollins, 2003. (A long, slow, imaginary look at life in the late 17th and early 18th centuries, with occasional peeks at Leibniz, Newton, and their quarrel.)

Westfall, Richard S. *Never at Rest: A Biography of Isaac Newton*. New York: Cambridge University Press, 1980.

Wilson, Grove. "Isaac Newton." Chapter 18 in *The Human Side of Science* (New York: Cosmopolitan Book Corp., 1929), pp. 189–207. An old-fashioned panegyric. Fun to read.

Youschkevitch, A. P. "Newton, Isaac." In Gillispie, vol. 10, 1974, pp. 42–103.

4. Bernoulli versus Bernoulli

Bernoulli. *The Bernoulli Edition: The Collected Scientific Papers of the Mathematicians and Physicists of the Bernoulli Family*. Patricia Radelet-de Grave, general editor. (Multivolume set of the collected works and correspondence of nine members of the family. All texts are printed in the original French and Latin; annotations are mostly in English. Basel: Birkhäuser, various dates.)

Bernoulli, Jacques (Jakob). *On the Theory of Combinations* (selection). In Smith, 1959, pp. 272–77.

Bernoulli, Jakob. "The Law of Large Numbers." From *Ars Conjectandi* (1713). In Newman, vol. 3, 1956, pp. 1452–54.

Bernoulli, Jean (Johann, John). "On the Brachistochrone Problem." Selections from *Acta Eruditorum* (June 1696, January 1697, and May 1697). In Smith, 1959, pp. 644–55. (Also includes some commentary by the editor.)

Bernoulli, Johann. "From 'The Curvature of a Ray in Nonuniform Media (1697) (The Brachistichrone).'" In Calinger, 1995, pp. 426–28.

Boyer, Carl B. "The First Calculus Textbooks." Chapter 80 in Swetz, 1994, pp. 532–39.

Calinger, Ronald. "The Bernoullis: Jakob Bernoulli." In Calinger, 1995, pp. 418–20.

———. "Johann Bernoulli." In Calinger, 1995, pp. 424–25.

Dunham, William. "The Bernoullis and the Harmonic Series." In Swetz, 1994, pp. 527–31.

———. "Bernoulli Trials." In Dunham, 1994, pp. 11–22.

Eves, Howard. "The Bernoulli Family." In *An Introduction to the History of Mathematics*. New York: Holt, Rinehart and Winston, rev. ed., 1964, pp. 355–58.

Fellman, E. A., and J. O. Fleckenstein. "Bernoulli, Johann (Jean) I." In Gillispie, vol. 2, 1970, pp. 51–55.

Gonzales, Tina. "Family Squabbles: The Bernoulli Family." Online: www.math.wichita .edu/history/men/bernoulli.html. November 24, 2003.

Guillen, Michael. *Five Equations That Changed the World*. New York: Hyperion, 1995. Especially section on the Bernoullis (pp. 65–117).

Hald, Anders. *A History of Probability and Statistics and Their Applications before 1750*. Hoboken, N.J.: John Wiley & Sons, 2003. (Orig. pub. 1990.)

Hofmann, J. E. "Bernoulli, Jakob (Jacques) I." In Gillispie, vol. 2, 1970, pp. 46–51.

Katz, V. J. "Analysis in the Eighteenth Century." In Katz, 1993, pp. 494–525.

Menger, Karl. "What Is Calculus of Variations and What Are Its Applications?" In Newman, vol. 2, 1956, pp. 886–90.

Newman, James R. "Commentary on the Bernoullis." In Newman, vol. 2, 1956, pp. 771–73.

O'Connor, J. J., and E. F. Robertson. "The Brachistochrone Problem." Online: www-history.mcs.st-andrews.ac.uk/HistTopics/Brachistochrone.html. February 2002.

———. "Daniel Bernoulli." Online: www-history.mcs.st-andrews.ac.uk/history/ Mathematicians/Bernoulli_Daniel.html. September 1998.

———. "Jacob (Jacques) Bernoulli." Online: www-history.mcs.st-andrews.ac.uk/history/ Mathematicians/Bernoulli_Jacob.html. September 1998.

———. "Johann Bernoulli." Online: www-history.mcs.st-andrews.ac.uk/history/ Mathematicians/Bernoulli_Johann.html. September 1998.

Rusting, Ricki. "Hatchet Job. A Late but Peaceful End to a Long and Nasty Dispute." (Johann and Brook Taylor). *Scientific American* (September 1990): 34.

Sensenbaugh, Roger. "The Bernoulli Family." *Great Lives from History: Renaissance to 1900 Series*, Frank N. Magill, ed., vol. 1, pp. 185–89. (Pasadena, Calif.: Salem Press, 1989).

Sierksma, Gerard. "Johann Bernoulli (1667–1748): His Ten Turbulent Years in Groningen." *Mathematical Intelligencer* 14, no. 4 (1992): 22–31.

Speiser, David. "The Bernoullis in Basel." *Mathematical Intelligencer* 14, no. 4 (1992): 46–47. Some background information on where the Bernoullis lived and worked in Basel.

Straub, Hans. "Bernoulli, Daniel." In Gillispie, vol. 2, 1970, pp. 36–46.

Thiele, Rüdiger. "Das Zerwürfnis Johann Bernoullis mit seinem Bruder Jakob" ("The Battle between Johann Bernoulli and His Brother Jakob"). *Acta historica Leopoldina* 27 (1997): 257–76.

Wikipedia Encyclopedia. Online: en.wikipedia.org/wiki/Daniel_Bernoulli. "Daniel Bernoulli," pp. 1, 2. Accessed September 21, 2004.

5. Sylvester versus Huxley

Ashforth, Albert. *Thomas Henry Huxley*. New York: Twayne, 1969.

Bell, E. T. "Invariant Twins, Cayley and Sylvester." Chapter 21 in Bell, 1937, pp. 378–405.

Bibby, Cyril, ed. *The Essence of T. H. Huxley: Selections from His Writings*. New York: St.

Martin's, 1967. (Not much here. Some background [1880s] on, for example, science and scientific method; also philosophy and psychology.)

——. *Scientist Extraordinary: The Life and Scientific Work of Thomas Henry Huxley 1825–1895.* New York: St. Martin's, 1972.

Cantor, Geoffrey. "Creating the Royal Society's Sylvester Medal." *British Journal for the History of Science* 37, no. 132 (March 2004): 75–92.

Desmond, Adrian. *Huxley: From Devil's Disciple to Evolution's High Priest.* Reading, Mass.: Addison-Wesley, 1997. (Orig. pub. 1994.)

Gill, Theodor Nicholas. "Thomas Henry Huxley." In Jones, 1966, pp. 491–513. (Some interesting background information.)

Grattan-Guinness, I. "The Contributions of J. J. Sylvester, F.R.S., to Mechanics and Mathematical Physics." *Notes and Records of the Royal Society of London* 55, no. 2 (2001): 253–65.

Hellman, Hal. *Great Feuds in Science.* New York: John Wiley & Sons, 1998.

Henfrey, Arthur, and Thomas Henry Huxley, eds. *Scientific Memoirs [of Thomas Henry Huxley].* New York: Johnson Reprint Co., 1966. (Orig. pub. 1853.) Also, Michael Foster and E. Ray Lankester. *The Scientific Memoirs of Thomas Henry Huxley,* 4 vols., suppl. London: Macmillan, 1898–1903.

Huxley, Thomas Henry. "Address on University Education" (1876) In *Collected Essays* III, pp. 235–261.

——. *Collected Essays by T. H. Huxley.* Nine volumes, especially vol. 1, *Method and Results*; vol. 3, *Science and Education*; and vol. 8, *Discourses: Biological & Geological.* New York: D. Appleton, 1899.

——. "Geological Reform" (Presidential Address to the Geological Society [1869]). In *Collected Essays* VIII, pp. 305–39.

——. *Life and Letters of Thomas Henry Huxley.* Two volumes. By his son, Leonard Huxley. New York: D. Appleton, 1900.

——. "On the Educational Value of the Natural History Sciences" (1854). In *Collected Essays* III, pp. 38–65.

——. "On Natural History, as Knowledge, Discipline, and Power." *Proceedings of the Royal Institution* (1856). In *Scientific Memoirs,* 1966, pp. 305–14.

——. "On the Physical Basis of Life" (based on an address delivered in Edinburgh on the evening of Sunday, the 8th of November, 1868). In *Collected Essays* I, pp. 130–65.

——. "On Science and Art in Relation to Education" (1882). (An address to the members of the Liverpool Institution.) In *Collected Essays* III, pp. 160–88.

——. "The Scientific Aspects of Positivism." *Fortnightly Review* 5 (1869): 653–70. In *Lay Sermons, Addresses and Reviews.* London (1870), pp. 128–50.

——. "Scientific Education: Notes of an After-Dinner Speech," *Macmillan's Magazine* 20 (June 1869): 177–84. In *Collected Essays* III, pp. 111–33.

——. "Universities: Actual and Ideal" (1874). In *Collected Essays* III, pp. 189–234.

Kenschaft, Patricia C., and Kaila Katz. "Sylvester and Scott." In Swetz, 1994, pp. 650–52.

Macfarlane, Alexander. "James Joseph Sylvester." Lecture delivered 1902. In Macfarlane, 1916, pp. 107–21.

Midonick, H. "James Joseph Sylvester," plus selections from Sylvester's 1869 Presidential Address to the British Association and 1877 Commemoration Day Address at Johns Hopkins University. In Midonick, 1965, pp. 759–73.

Paradis, James G. *T. H. Huxley: Man's Place in Nature.* Lincoln: University of Nebraska Press, 1978.

Parshall, Karen Hunger. *James Joseph Sylvester: Jewish Mathematician in a Victorian World*. Baltimore: Johns Hopkins University Press, 2006.

——. *James Joseph Sylvester: Life and Work in Letters*. Oxford, England: Clarendon, 1998.

Sylvester, James Joseph. "Presidential Address to Section 'A' of the British Association" (1869). In *The Collected Mathematical Papers of James Joseph Sylvester*, vol. 2 (1854–1873). Cambridge, England: Cambridge University Press, 1908, pp. 650–61.

Williams, Wesley C. "Huxley, Thomas Henry." In Gillispie, vol. 6, 1972, pp. 589–97.

6. Kronecker versus Cantor

Aczel, Amir D. *The Mystery of the Aleph: Mathematics, the Kabbalah, and the Search for Infinity*. New York: Four Walls Eight Windows, 2000.

Barrow, John D. "The Tragedy of Cantor and Kronecker." In Barrow, 1992, pp. 198–216.

Biermann, Kurt-R. "Kronecker, Leopold." In Gillispie, vol. 7, 1973, pp. 505–9.

Burton, David M. "Counting the Infinite." In Burton, 1991, pp. 624–46.

——. "The Paradoxes of Set Theory." In Burton, 1991, pp. 647–63.

Chaitin, G. J. "A Century of Controversy over the Foundations of Mathematics." Lecture at Carnegie Mellon University, March 2, 2000. Online: http://www.umcs.maine.edu/~chaitin/cmu.pdf.

Dauben, Joseph W. "Georg Cantor and the Origins of Transfinite Set Theory." *Scientific American* 248 (June 1983): 122–31.

Dauben, J. W. *Georg Cantor: His Mathematics and Philosophy of the Infinite*. Princeton, N.J.: Princeton University Press, 1990; Cambridge, Mass.: Harvard University Press, 1979.

Dunham, William. "The Non-Denumerability of the Continuum (1874)," and "Cantor and the Transfinite Realm (1891)." Chapters 11 and 12 in Dunham, 1990, pp. 245–83.

Edwards, Howard. "An Appreciation of Kronecker." *Mathematical Intelligencer* 9, no. 1 (1987): 28–35.

——. "Kronecker's Place in History." In Aspray and Kitcher, 1988, pp. 139–44.

Gardner, Martin. "Mathematical Games: The Hierarchy of Infinities and the Problems It Spawns." *Scientific American* 214, no. 3 (March 1966): 112–18.

Goldstein, Rebecca. *Incompleteness: The Proof and Paradox of Kurt Gödel*. New York: W. W. Norton, 2005.

Grattan-Guinness, Ivor. "Towards a Biography of Georg Cantor." *Annals of Science* 27, no. 4 (1971a): 345–91.

Henderson, Harry. "Georg Cantor (1845–1918)." In *Modern Mathematicians*. New York: Facts On File, 1996, pp. 26–35.

Hill, Clair Ortiz. "Frege's Attack on Husserl and Cantor." *Monist* 77, no. 3 (July 1994): 345–57.

Jones, Phillip S. "Irrationals or Incommensurables V: Their Admission to the Realm of Numbers." In Swetz, 1994, pp. 668–70.

Kanamori, Akihiro. "The Mathematical Import of Zermelo's Well-Ordering Theorem." *The Bulletin of Symbolic Logic* 3, no. 3 (September 1997): 281–311.

Kaplan, Robert and Ellen. *The Art of the Infinite*. New York: Oxford University Press, 2003.

Klarreich, Erica. "Infinite Wisdom." *Science News* 164 (August 30, 2003): 139–41.

Laubenbacher and Pangelley. "Set Theory: Taming the Infinite." Chapter 2 in *Mathematical Expeditions: Chronicles by the Explorers*, 1999, pp. 54–89.

Love, William P. "Infinity: The Twilight Zone of Mathematics." Chapter 101 in Swetz, 1994, pp. 658–67.

Meschkowski, H. "Cantor, Georg." In Gillispie, vol. 3, 1971, pp. 52–58.

O'Connor, J. J., and E. F. Robertson. "Georg Cantor." Online: www-groups.dcs .st-and.ac.uk/~history/Mathematicians/Cantor.html, 1998.

———. "A History of Set Theory." Online: www-groups.dcs.st-and.ac.uk/~history/ HistTopics/Beginnings_of_set_theory.html, 1996.

———. "Leopold Kronecker." Online: http://www-groups.dcs.st-and.ac.uk/~history/ Mathematicians/Kronecker.html, 1999.

Pappas, Theoni. "Cantor Driven to Nervous Breakdown." In Pappas, 1997, pp. 50–58.

Romano, Carlin. "Godel and Cantor; Madness as Accessory to Genius." *Chronicle of Higher Education* 47, no. 2 (September 8, 2000): B10.

Rucker, Rudy. *Infinity and the Mind: The Science and Philosophy of the Infinite.* Boston: Birkhâuser, 1982.

Sweeney, Seamus. "The Mystery of Infinity." *Lancet* 357, no. 9255 (February 17, 2001): 564–65. (Essay review of Aczel, 2000.)

Wallace, David Foster. *Everything and More: A Compact History of* ∞ . New York: W.W. Norton, 2003.

7. Borel versus Zermelo

Abbott, David. "Borel, Émile Félix-Édouard-Justin." In Abbott, 1986, p. 22.

———. "Zermelo, Ernst Friedrich Ferdinand." In Abbott, 1986, pp. 142–43.

Aczel, Amir D. "The Axiom of Choice." Chapter 14 in Azcel, 2000, pp. 171–78.

Bell, Eric Temple. Bell's take on the "spirited controversy" that was touched off by Zermelo's "notorious" axiom. Bell, 1945, pp. 484–85.

Calinger, Ronald. "Ernst (Friedrich Ferdinand) Zermelo." In Calinger, 1995, pp. 719–20.

Cohen, Paul I., and Reuben Hersh. "Non-Cantorian Set Theory." *Scientific American* (December 1967): 104–16.

Dauben, Joseph. "Zermelo, Well-Ordering, and the Axiom of Choice." In Dauben, 1990, pp. 250–59.

Heinzmann, Gerhard. *Poincaré, Russell, Zermelo et Peano. Textes de la discussion (1906–1912) sur les fondements des mathématiques: des antinomies à la prédicativité.* Paris: Librairie Scientifique et Technique, 1986.

Hirsch, Morris W. "Book Review." *Bulletin of the American Mathematical Society* 32, no. 1 (January 1995): 137–48. (Essay review of *Realism in Mathematics*, by Penelope Maddy, 1993).

Irvine, A. D. "Russell's Paradox." 2003 (*Stanford Encyclopedia of Philosophy*), pp. 1–4. Online: www.science.uva.nl/~seop/entries/russell-paradox/.

Jervell, Herman R. "From the Axiom of Choice to Choice Sequences." *Nordic Journal of Philosophical Logic* 1, no. 1 (1996): 95–98.

Maddy, Penelope. *Realism in Mathematics.* New York: Oxford University Press, 1990.

May, Kenneth D. "Borel, Émile (Félix Édouard-Justin)." In Gillispie, vol. 2, 1970, pp. 302–5.

Moore, Gregory H. *Zermelo's Axiom of Choice: Its Origins, Development, and Influence.* New York: Springer-Verlag, 1982.

O'Connor, J.J. and E. F. Robertson. "Ernst Friedrich Ferdinand Zermelo." Online, 1999, pp. 1–3: www-groups.dcs.st-and.ac.uk/~history/Mathematicians/Zermelo.html.

———. "A History of Set Theory." Online, February 1996: www-groups.dcs.st-and.ac .uk/~history/HistTopics/Beginnings_of_set_theory.html.

Russell, Bertrand. "On Some Difficulties in the Theory of Transfinite Numbers and Order Types" (December 14, 1906), in Heinzmann, 1986, pp. 54–78.

Schechter, Eric. "A Home Page for the Axiom of Choice." Online, updated March 7, 2005: math.vanderbilt.edu/~schectex/ccc/choice.html.

Van Rootselaar, B. "Zermelo, Ernst Friedrich Ferdinand." In Gillispie, vol. 14, 1976, pp. 613–16.

Whitehead, Alfred North, and Bertrand Russell. *Principia Mathematica*. (Abridged text of vol. 1., 1962. Orig. pub. 1910.) New York: Cambridge University Press, 1997.

Zermelo, Ernst. "Proof That Every Set Can Be Well-Ordered." From a letter to David Hilbert (September 24, 1904). In Calinger, 1995, pp. 720–22.

8. Poincaré versus Russell

Boole, George. "Mathematical Analysis of Logic." In Newman, vol. 3, 1956, pp. 1856–58.

Broadbent, T. A. A. "Russell, Bertrand Arthur William." In Gillispie, vol. 12, 1975, pp. 9–17.

Detlefsen, Michael. "Poincaré against the Logicians." *Synthese* 90 (1992): 349–78.

———. "Poincaré vs Russell on the Role of Logic in Mathematics." *Philosophia Mathematica* 1, no. 3 (1993): 24–49.

Dieudonné, Jean. "Poincaré, Jules Henri." In Gillispie, vol. 11, 1975, pp. 51–61.

Durant, Will. "Bertrand Russell." In *The Story of Philosophy*. New York: Simon & Schuster, 1933, pp. 357–64. (Orig. pub. 1926.)

Galison, Peter. *Einstein's Clocks, Poincaré's Maps*. New York: W. W. Norton, 2003.

Goldfarb, Warren. "Poincaré against the Logicists." In Aspray and Kitcher, 1988, pp. 61–81.

Gray, Jeremy. "The Shock of the New Maths." *New Statesman & Society* 6, no. 282 (1993): 30–31.

Irvine, Andrew. "Bertrand Russell." Stanford Encyclopedia of Philosophy (online). http://plato.stanford.edu/entries/russell/, 2003a.

———. "Principia Mathematica." Stanford Encyclopedia of Philosophy (online). http://plato.stanford.edu/entries/principia-mathematica/, 2003b.

Lalande, André. "Henri Poincaré: From Science and Hypothesis to Last Thoughts." In Wiener, 1957, pp. 624–26.

Lambek, Jim. "Are the Traditional Philosophies of Mathematics Really Incompatible?" *Mathematical Intelligencer* 16, no. 1 (1994): 56–62.

Linsky, Bernard, and Edward N. Zalta. "What Is Neologicism?" Online at http://mally.stanford.edu/neologicism2.pdf, 2004, pp. 1–37.

Moorhead, Caroline. *Bertrand Russell: A Life*. New York: Viking, 1992.

Newman, James R. "Commentary on an Absent-Minded Genius and the Laws of Chance" (Poincaré). In Newman, vol. 2, 1956, pp. 1374–79.

Nordman, Charles. "Jules Henri Poincaré." In Jones, 1966, pp. 615–37. (Some interesting background information.)

O'Connor, J. J., and E. F. Robertson. "Bertrand Arthur William Russell." University of St. Andrews online (www-history.mcs.st-andrews.ac.uk/Mathematicians/Russell.html), 1996.

———. "Friedrich Ludwig Gottlob Frege." University of St. Andrews online (www-history.mcs.st-andrews.ac.uk/Mathematicians/Frege.html), 2002.

———. "Jules Henri Poincaré." University of St. Andrews online (www-history.mcs.st-andrews.ac.uk/Mathematicians/Poincaré.html), 2003.

Poincaré, Henri. *The Foundations of Science*, trans. George Bruce Halsted. Lancaster, Pa.:
    Science Press, 1946. (Orig. pub. 1913.) (Especially portions of Book 2, "Mathemat-
    ical Reasoning," in "Science and Method," pp. 448–85 [see Poincaré, May 1906].)
——. "La logique de l'infini." *Revue de Métaphysique et de Morale* 17 (1909): 461–82.
——. "Mathematical Creation" (1908). In Newman, vol. 4, 1956, pp. 2041–50.
——. "Mathematical Discovery." In Rapport and Wright, 1964, pp. 128–37.
——. "Les mathématiques et la logique." *Revue de Métaphysique et de Morale* (part 1) 13
    (1905): 815–35; (part 2) 14 (May 1906): 17–34. English translation by George Bruce
    Halsted, in Poincaré, 1946 (orig. pub. 1913), chapter 3, "Mathematics and Logic,"
    pp. 448–59; chapter 4, "The New Logics," pp. 460–71; and chapter 5, "The Latest
    Efforts of the Logisticians," pp. 472–85 (all in Book 2 of the section called "Science
    and Method" of *The Foundations of Science*).
Russell, Bertrand. *The Autobiography of Bertrand Russell, 1872–1914; 1914–1944*. Boston:
    Little, Brown, vol. 1, 1967; vol. 2, 1968.
——. *Essays in Analysis*. Ed. Douglas Lackey. New York: George Braziller, 1973.
——. *Introduction to Mathematical Philosophy*. New York: Routledge, 1993.(Orig. pub. 1919.)
——. "Mathematical Logic As Based on the Theory of Types." In Russell, *Logic and
    Knowledge. Essays 1901–1950*, ed. Robert Charles Marsh. New York: Macmillan,
    1956. (Orig. pub. 1908.)
——. "Mathematics and the Metaphysicians" (1929). In Newman, vol. 3, 1956, pp.
    1576–90.
——. "My Mental Development" (1954). In Newman, vol. 1, 1956, pp. 381–94.
——. *My Philosophical Development*. New York: Simon and Schuster, 1959.
——. *Mysticism and Logic* (essays). New York: Doubleday, undated (1958?).
——. *Portraits from Memory*. New York: Simon and Schuster, 1956.
——. *Principles of Mathematics*. 2nd ed. New York: W. W. Norton, 1938. (Orig. pub. 1903.)
——. "On Some Difficulties in the Theory of Transfinite Numbers and Order Types."
    Proceedings of the London Mathematical Society, series 2, 4 (March 7, 1906): 29–53. (In
    Heinzmann, 1986, pp. 54–78. Also chapter 7 in Russell, 1973, pp. 135–64.)
——. "The Study of Mathematics" (1902). In Russell, 1917, pp. 55–69.
——. "La Théorie des Types Logiques." *Revue de Métaphysique et de Morale* 18 (May
    1910): 263–301. English translation: "The Theory of Logical Types," chapter 10 in
    Russell, 1973, pp. 215–52.
Slater, John G. *Bertrand Russell*. Bristol, England: Thoemmes Press, 1994.
Snapper, Ernst. "The Three Crises in Mathematics: Logicism, Intuitionism, and Formal-
    ism." Chapter 106 in Swetz, 1994, pp. 697–707.
Stein, Howard. "Logos, *Logic*, and Logistiké: Some Philosophical Remarks on
    Nineteenth-Century Transformation of Mathematics." In Aspray and Kitcher,
    1988, pp. 238–59. ("Progenitors or foreshadowers" of logicism, formalism, and
    intuitionism.)
Tenn, Joseph S. "Henri Poincaré: The Ninth Bruce Medalist." *Mercury* 20, no. 4
    (July/August 1991): 111–12.
Schuyler. "Russell, Bertrand. Originator of Logicism." In Simonis, 1999, p. 172.
Van Evra, James W. "The Search for Mathematical Roots, 1870–1940: Logistics, Set The-
    ories, and the Foundations of Mathematics from Cantor through Russell to Godel."
    Isis 94, no. 2 (June 2003): 387. Review of I. Grattan-Guinness's book by this name.
Van Heijenoort, J., ed. *From Frege to Gödel: a Source Book in Mathematical Logic, 1879–1931*.
    Cambridge, Mass.: Harvard University Press, 1967.
Young, Robyn V. "Bertrand Russell." In Young, 1998, pp. 428–31.

9. Hilbert versus Brouwer

Barrow, John D. "The Comedy of Hilbert and Brouwer." In Barrow, 1992, pp. 216–26.

Brouwer, Luitzen Egbertus Jan. "Intuitionist Set Theory." Paper presented to the Royal Academy of Sciences, December 18, 1920. In Mancosu, 1998, pp. 23–39.

Hatcher, Donald L. "Epistemology and Education: A Case for Fallibilism." Online, undated: www.bakeru.edu/crit/literature/dlh_ct_epistemology.htm.

Hays, John. "The Battle of the Frog and the Mouse (from the Fables of Aleph)." *Mathematical Intelligencer* 6, no. 2 (1984): 77–80.

Hilbert, David. "From 'Mathematical Problems: Lecture Delivered before the International Congress of Mathematicians at Paris in 1900. (Paris Problems and the Formalist Program).'" In Calinger, 1995, pp. 698–718.

Mancosu, Paolo. *From Brouwer to Hilbert.* New York: Oxford University Press, 1998.

Nyikos, Peter. "Hilbert's First and Second Problems and the Foundations of Mathematics." *Topology Atlas Invited Contributions* 9, no. 3 (2004): 1–6. Online: PDF form is http://arxiv.org/PS_cache/math/pdf/0412/0412555.pdf.

Peckhaus, Volker. "The Pragmatism of Hilbert's Program." *Synthese* 137 (2003): 141–56. Online, pp. 1–17: www-fakkw.upb.de/institute/philosophie/Personal/Peckhaus/Texte_zum_Download/pragmatism-h.pdf.

Reid, C. *Hilbert.* New York: Springer-Verlag, 1970.

Simpson, Stephen G. "Partial Realizations of Hilbert's Program," February 4, 1986, pp. 1–22. Online: ftp://ftp.math.psu.edu/simpson/papers/hilbert.pdf. Published in the *Journal of Symbolic Logic* 53 (1988): 349–63.

Snapper, Ernst. "The Three Crises in Mathematics: Logicism, Intuitionism, and Formalism." Chapter 106 in Swetz, 1994, pp. 697–707.

Stein, Howard. "Logos, Logic, and Logistiké: Some Philosophical Remarks on Nineteenth-Century Transformation of Mathematics." In Aspray and Kitcher, 1988, pp. 238–59.

van Dalen, Dirk. *Mystic, Geometer, and Intuitionist: The Life of L. E. J. Brouwer.* New York: Oxford University Press. Vol. 1, The Dawning Revolution, 1999. Vol. 2, *Hope and Disillusion,* 2005.

———. "The War of the Frogs and the Mice, or the Crisis of the *Mathematische Annalen.*" *Mathematical Intelligencer* 12, no. 4 (1990): 17–31.

van Stigt, Walter P. *Brouwer's Intuitionism.* New York: Elsevier Science, 1990.

———. "Brouwer's Intuitionist Program." In Mancosu, 1998, pp. 1–22.

von Mises, Richard. "Logistic" and "The Foundations of Mathematics." Sections in "Mathematical Postulates and Human Understanding" (1951). In Newman, vol. 3, 1956, pp. 1733–54.

Weyl, Hermann. "David Hilbert and His Mathematical Work" (1944). In Reid, 1970, pp. 245–83. (Especially Axiomatics, pp. 264–74).

Young, Robyn V. "David Hilbert." In Young, 1998, pp. 244–47.

———. "Luitzen Egbertus Jan Brouwer." In Young, 1998, p. 81.

10. Absolutists/Platonists versus Fallibilists/Constructivists

Barnard, Tony, and Peter Saunders. "Superior Sums That Don't Add Up to Much. How Mathematics Is Taught Is in Dispute, and Our Children's Knowledge of It Is Worse Than Ever." *The Guardian* (December 28, 1994), Features, p. 18.

Barrow, John D. "Platonic Heavens above and Within." Chapter 6 in Barrow, 1992, pp. 249–97.

Ernest, Paul. "Education: Number Crunch; Too Much Maths Teaching Is Outdated and Boring." *The Guardian* (January 3, 1995), Education Page, p. T15.

——. "In response to Professor Zheng." *Philosophy of Mathematics Education Newsletter* 7 (February 1994), p. 5. Online: www.people.ex.ac.uk/PErnest/pome/pome7.htm.

——. "Is Mathematics Discovered or Invented?" *Philosophy of Mathematics Education Journal 12* (November 1999). Online: www.people.ex.ac.uk/PErnest/pome12/article2.htm.

——. "What Is the Philosophy of Mathematics Education?" *Philosophy of Mathematics Education Journal 18* (October 2004). Online: www.people.ex.ac.uk/PErnest/pome18/PhoM_%20for_ICME_04.htm.

Hardy, G. H. *A Mathematician's Apology.* London: Cambridge University Press, 1967. (Orig. pub. 1941.)

Hersh, Reuben. "Some Proposals for Reviving the Philosophy of Mathematics." *Advances in Mathematics* 31 (1979): 31–50.

Kline, Morris. *Mathematics: The Loss of Certainty.* New York: Oxford University Press, 1980.

Macleod, Donald. "School Maths 'Is in Crisis.' Experts Accused of Complacency as Universities Demand Review." *The Guardian* (December 28, 1994), Home Page, p. 1.

Penrose, Roger. *The Emperor's New Mind: Concerning Computers, Minds and the Laws of Physics.* New York: Penguin Books USA, 1991. (Orig. pub. 1989.)

Pepin, Birgit. "Epistemologies, Beliefs and Conceptions of Mathematics Teaching and Learning: The Theory, and What Is Manifested in Mathematics Teachers' Work in England, France and Germany." *TNTEE Publications* 2, no. 1 (October 1999): 127–46.

Roulet, Geoffrey. "The Philosophy of Mathematics Education: 'What Does This Mean for the Child in the Classroom?'" *Philosophy of Mathematics Education Newsletter* 6 (December 1992). Online: www.people.ex.ac.uk/PErnest/pome/pome6.htm.

Rowlands, Stuart, and Robert Carson. "Contradictions in Constructivist Discourse." *Philosophy of Mathematics Education Journal* 14 (May 2001). Online: www.people.ex.ac.uk/PErnest/pome14/rowlands.htm.

Schoenfeld, Alan H. "Making Mathematics Work for All Children: Issues of Standards, Testing, and Equality." *Educational Researcher* 31, no. 1 (January/February 2002): 13–25.

Thom, René. "Why the Philosophy of Mathematics Education?" *Philosophy of Mathematics Education Newsletter*, no. 1 (July 1990). Online: www.people.ex.ac.uk/PErnest/pome/pome1.htm.

Wu, H. "The Mathematician and the Mathematics Education Reform." *Notices of the AMS* 43 (December 1996): 1531–37.

Index